NUREG-1949, Vol. 1

Safety Evaluation Report Related to Disposal of High-Level Radioactive Wastes in a Geologic Repository at Yucca Mountain, Nevada

Volume 1: General Information

Manuscript Completed: August 2010
Date Published: August 2010

Office of Nuclear Material Safety and Safeguards

NOTE TO READER: This volume is one of five volumes shown in the table below that comprise the Safety Evaluation Report (SER). Each volume is to be published separately as it is completed; however, the volume number may not be published in sequence (e.g., Volume 3 is anticipated to be published before Volume 2). The SER volume number and section number of chapters within a volume, are based on the Yucca Mountain Review Plan (YMRP)[1] that the U.S. Nuclear Regulatory Commission (NRC) staff used to guide its review of the license application. Use of SER section numbers that correspond to the YMRP section numbers facilitated the NRC staff's writing of the SER, allowing the interested reader to easily find the applicable review methods and acceptance criteria within the YMRP. The following table provides the topics and SER sections for each volume. The table will help the reader locate SER section cross-references in each volume.

Chapter	SER Section	Title
Volume 1 General Information		
1	1.1	General Description
2	1.2	Proposed Schedules for Construction, Receipt, and Emplacement of Waste
3	1.3	Physical Protection Plan
4	1.4	Material Control and Accounting Program
5	1.5	Description of Site Characterization Work
Volume 2 Repository Safety Before Permanent Closure		
1	2.1.1.1	Site Description as it Pertains to Preclosure Safety Analysis
2	2.1.1.2	Description of Structures, Systems, Components, Equipment, and Operational Process Activities
3	2.1.1.3	Identification of Hazards and Initiating Events
4	2.1.1.4	Identification of Event Sequences
5	2.1.1.5	Consequence Analyses
6	2.1.1.6	Identification of Structures, Systems, and Components Important to Safety; and Measures to Ensure Availability of the Safety Systems
7	2.1.1.7	Design of Structures, Systems, and Components Important to Safety and Safety Controls
8	2.1.1.8	Meeting the 10 CFR Part 20 As Low As Is Reasonably Achievable Requirements for Normal Operations and Category 1 Event Sequences
9	2.1.2	Plans for Retrieval and Alternate Storage of Radioactive Wastes
10	2.1.3	Plans for Permanent Closure and Decontamination, or Decontamination and Dismantlement of Surface Facilities
Volume 3 Repository Safety After Permanent Closure		
1	2.2.1.1	System Description and Demonstration of Multiple Barriers
2	2.2.1.2.1	Scenario Analysis
3	2.2.1.2.2	Identification of Events with Probabilities Greater Than 10^{-8} Per Year
4	2.2.1.3.1	Degradation of Engineered Barriers
5	2.2.1.3.2	Mechanical Disruption of Engineered Barriers
6	2.2.1.3.3	Quantity and Chemistry of Water Contacting Engineered Barriers and Waste Forms

---

[1]NRC. 2003. NUREG–1804, "Yucca Mountain Review Plan—Final Report." Rev. 2. Washington, DC: NRC.

Chapter	SER Section	Title
\multicolumn{3}{l}{**Volume 3 Repository Safety After Permanent Closure (continued)**}		
7	2.2.1.3.4	Radionuclide Release Rates and Solubility Limits
8	2.2.1.3.5	Climate and Infiltration
9	2.2.1.3.6	Unsaturated Zone Flow
10	2.2.1.3.7	Radionuclide Transport in the Unsaturated Zone
11	2.2.1.3.8	Flow Paths in the Saturated Zone
12	2.2.1.3.9	Radionuclide Transport in the Saturated Zone
13	2.2.1.3.10	Igneous Disruption of Waste Packages
14	2.2.1.3.12	Concentration of Radionuclides in Groundwater
15	2.2.1.3.13	Airborne Transportation and Redistribution of Radionuclides*
16	2.2.1.3.14	Biosphere Characteristics
17	2.2.1.4.1	Demonstration of Compliance with the Postclosure Individual Protection Standard
18	2.2.1.4.2	Demonstration of Compliance with the Human Intrusion Standard
19	2.2.1.4.3	Demonstration of Compliance with the Separate Groundwater Protection Standards
20	2.5.4	Expert Elicitation
\multicolumn{3}{l}{**Volume 4 Administrative and Programmatic Requirements**}		
1	2.3	Research and Development Program to Resolve Safety Questions
2	2.4	Performance Confirmation Program
3	2.5.1	Quality Assurance Program
4	2.5.2	Records, Reports, Tests, and Inspections
5	2.5.3.1	U.S. Department of Energy Organizational Structure as it Pertains to Construction and Operation of Geologic Repository Operations Area
6	2.5.3.2	Key Positions Assigned Responsibility for Safety and Operations of Geologic Repository Operations Area
7	2.5.3.3	Personnel Qualifications and Training Requirements
8	2.5.5	Plans for Startup Activities and Testing
9	2.5.6	Plans for Conduct of Normal Activities, Including Maintenance, Surveillance, and Periodic Testing
10	2.5.7	Emergency Planning
11	2.5.8	Controls to Restrict Access and Regulate Land Uses
12	2.5.9	Uses of Geologic Repository Operations Area for Purposes Other Than Disposal of Radioactive Wastes
\multicolumn{3}{l}{**Volume 5 License Specifications**}		
1	2.5.10	License Specifications

*This SER Section combines the review of information addressed in the YMRP Sections 2.2.1.3.11 and 2.2.1.3.13.

ABSTRACT

This is the first volume of the U.S. Nuclear Regulatory Commission (NRC) staff's "Safety Evaluation Report Related to Disposal of High-Level Radioactive Wastes in a Geologic Repository at Yucca Mountain, Nevada." It documents the NRC staff's review and evaluation of general information the U.S. Department of Energy (DOE) provided in its June 3, 2008, license application, as updated on February 19, 2009, that seeks an authorization to begin construction of a repository at Yucca Mountain. In subsequent volumes of the report, Volumes 2–5, the NRC staff plans to present its review and evaluation of the Safety Analysis Report included in DOE's license application.

Consistent with NRC's requirements for the general information, the NRC staff reviewed the following: (i) a general description of the proposed repository, (ii) proposed schedules for repository activities, (iii) a description of security measures, (iv) a description of the Material Control and Accounting Program, and (v) a description of work done to characterize the site.

On the basis of its review and specified DOE commitments, the NRC staff concludes in this volume that DOE has provided information that satisfies the requirements of 10 CFR 63.21(b)(1)-(5) of the NRC's regulations.

CONTENTS

Section	Page
ABSTRACT	v
EXECUTIVE SUMMARY	xi
ACRONYMS AND ABBREVIATIONS	xv
INTRODUCTION	1

CHAPTER 1 1-1
  1.1 General Description 1-1
    1.1.1   Introduction 1-1
    1.1.2   Regulatory Requirements 1-1
    1.1.3   Technical Review 1-2
        1.1.3.1   Summary of the DOE License Application on General Description 1-2
        1.1.3.2   NRC Staff Evaluation of General Description 1-2
            1.1.3.2.1   Location and Arrangement of the GROA 1-2
            1.1.3.2.2   General Nature of the Activities To Be Conducted at the GROA 1-6
            1.1.3.2.3   Basis for the Exercise of NRC Licensing Authority 1-9
            1.1.3.2.4   General References 1-9
    1.1.4   Evaluation Findings 1-10
    1.1.5   References 1-10

CHAPTER 2 2-1
  1.2 Proposed Schedules for Construction, Receipt, and Emplacement of Waste 2-1
    1.2.1   Introduction 2-1
    1.2.2   Regulatory Requirements 2-1
    1.2.3   Technical Review 2-1
        1.2.3.1   Summary of the DOE License Application on Proposed Schedules 2-1
        1.2.3.2   NRC Staff Evaluation of Proposed Schedules 2-2
    1.2.4   Evaluation Findings 2-3
    1.2.5   References 2-3

CHAPTER 3 3-1
  1.3 Physical Protection Plan 3-1
    1.3.1   Introduction 3-1
    1.3.2   Regulatory Requirements 3-1
    1.3.3   Technical Review 3-3
        1.3.3.1   Summary of the DOE License Application on Physical Protection Plan 3-3
        1.3.3.2   NRC Staff Evaluation of Physical Protection Plan 3-3
            1.3.3.2.1   Description and Schedule for Implementation 3-3
            1.3.3.2.2   General Performance Objectives 3-5
            1.3.3.2.3   Protection Goal 3-6
            1.3.3.2.4   Security Organization 3-7
            1.3.3.2.5   Physical Barrier Subsystems 3-10
            1.3.3.2.6   Access Control Subsystems and Procedures 3-11
            1.3.3.2.7   Detection, Surveillance, and Alarm Subsystems and Procedures 3-13

CONTENTS (continued)

Section	Page

        1.3.3.2.8   Communication Subsystems 3-15
        1.3.3.2.9   Equipment Operability and Compensatory Measures 3-16
        1.3.3.2.10 Contingency and Response Plans and Procedures 3-17
        1.3.3.2.11 Reporting of Safeguards Events 3-18
        1.3.3.2.12 Findings on Physical Protection Plan 3-19
  1.3.4   Evaluation Findings 3-20
  1.3.5   References 3-20

CHAPTER 4 4-1
1.4 Material Control and Accounting Program 4-1
  1.4.1   Introduction 4-1
  1.4.2   Regulatory Requirements 4-1
  1.4.3   Technical Review 4-2
      1.4.3.1   Summary of the DOE License Application on Material Control and Accounting Program 4-3
      1.4.3.2   NRC Staff Evaluation of Material Control and Accounting Program 4-3
            1.4.3.2.1   Material Balance, Inventory, and Record-Keeping Procedures 4-3
                    1.4.3.2.1.1 Material Control and Accounting Plan 4-3
                    1.4.3.2.1.2 Record Information 4-5
                    1.4.3.2.1.3 Physical Inventory 4-5
                    1.4.3.2.1.4 Quality of Physical Inventories 4-5
                    1.4.3.2.1.5 Procedures 4-6
                    1.4.3.2.1.6 Detection of Falsification of Data and Reports ... 4-6
                    1.4.3.2.1.7 Records Storage 4-6
                    1.4.3.2.1.8 Summary of NRC Staff's Evaluation of Material Balance, Inventory, and Record-Keeping Procedures 4-6
            1.4.3.2.2   Reports of Accidental Criticality or Loss of Special Nuclear Material 4-7
            1.4.3.2.3   Material Status Reports 4-8
            1.4.3.2.4   Nuclear Material Transfer Reports 4-8
            1.4.3.2.5   Findings on Material Control and Accounting 4-9
  1.4.4   Evaluation Findings 4-9
  1.4.5   References 4-9

CHAPTER 5 5-1
1.5 Description of Site Characterization Work 5-1
  1.5.1   Introduction 5-1
  1.5.2   Regulatory Requirements 5-1
  1.5.3   Technical Review 5-1
      1.5.3.1   Summary of the DOE License Application on Site Characterization 5-2
      1.5.3.2   NRC Staff Evaluation of Site Characterization 5-2
            1.5.3.2.1   Site Characterization Activities 5-2
                    1.5.3.2.1.1 Geology 5-3
                    1.5.3.2.1.2 Hydrology 5-3
                    1.5.3.2.1.3 Geochemistry 5-3

CONTENTS (continued)

Section	Page

     1.5.3.2.1.4 Geotechnical Properties and Conditions of the Host Rock 5-4
     1.5.3.2.1.5 Climatology, Meteorology, and Other Environmental Sciences 5-4
     1.5.3.2.1.6 Reference Biosphere 5-4
     1.5.3.2.1.7 Summary of NRC Staff's Evaluation of Site Characterization Activities 5-5
   1.5.3.2.2 Site Characterization Results 5-5
     1.5.3.2.2.1 Geology 5-6
     1.5.3.2.2.2 Hydrology 5-8
     1.5.3.2.2.3 Geochemistry 5-11
     1.5.3.2.2.4 Geotechnical Properties 5-12
     1.5.3.2.2.5 Climatology, Meteorology, and Other Environmental Information 5-14
     1.5.3.2.2.6 Reference Biosphere 5-15
     1.5.3.2.2.7 Summary of NRC Staff's Evaluation of Site Characterization Results 5-16
   1.5.3.2.3 Summary of NRC Staff's Evaluation of Site Characterization 5-16
 1.5.4 Evaluation Findings 5-16
 1.5.5 References 5-16

CHAPTER 6 6-1
 Conclusions 6-1

CHAPTER 7 7-1
 Glossary 7-1

APPENDIX—Commitments A–1

EXECUTIVE SUMMARY

On June 3, 2008, the U.S. Department of Energy (DOE) submitted a license application to the U.S. Nuclear Regulatory Commission (NRC) seeking an authorization to begin construction of a geologic repository for high-level radioactive waste (HLW) disposal at Yucca Mountain, Nevada.[1] The license application consists of general information and a Safety Analysis Report (SAR). This document, the NRC staff's Safety Evaluation Report (SER), Volume 1, presents the results of the NRC staff's review of the general information DOE provided in its June 3, 2008, license application and as updated on February 19, 2009.[2] The NRC staff also reviewed information DOE provided in response to NRC staff's requests for additional information. As appropriate, the SER provides specific citations to these additional information sources in the context of the NRC staff's review. In subsequent volumes, SER Volumes 2–5, the NRC staff plans to present the results of its safety review of the DOE SAR. Any NRC decision on whether to authorize construction of a geologic repository for high-level radioactive waste (HLW) disposal at Yucca Mountain, Nevada, will be made only after the NRC staff has completed all volumes of the SER. A decision to issue a construction authorization will not be effective until after the Commission has completed its review under 10 CFR 2.1023. In conducting its review of the entire license application, the NRC staff was guided by the review methods and acceptance criteria outlined in the Yucca Mountain Review Plan.[3]

The NRC staff reviewed the general information DOE provided. Consistent with the NRC's regulations, DOE provided the following in its general information: (i) a general description of the proposed repository, (ii) proposed schedules for repository activities, (iii) a description of the security measures for physical protection of waste, (iv) a description of the Material Control and Accounting Program, and (v) a description of work conducted to characterize the Yucca Mountain site. A summary of the review of these five requirements is provided below.

General Description of the Proposed Repository

NRC requires that the general description identify the location of the geologic repository operations area (GROA). The GROA is that part of the proposed repository, including both surface and subsurface areas, where waste is handled. NRC also requires that DOE describe the general character of proposed repository activities and discuss the basis for the exercise of the Commission's licensing authority. This description is largely informational in nature. More detailed discussions and descriptions are found in the SAR. On the basis of the NRC staff's review of the general information and other information submitted in support of the license application, the NRC staff finds that DOE presented an adequate general description of the geologic repository that identified the location of the GROA, discussed the general character of the proposed activities at the GROA, and presented an adequate description of the basis for the exercise of the Commission's licensing authority.

Proposed Schedules for Repository Activities

The NRC staff reviewed DOE's proposed schedules for construction, receipt of waste, and emplacement of wastes at the proposed GROA to assure that the schedules adequately

---

[1] DOE. 2008. DOE/RW–0573, "Yucca Mountain Repository License Application." Rev. 0. ML081560400. Las Vegas, Nevada: DOE, Office of Civilian Radioactive Waste Management.
[2] DOE. 2009. DOE/RW–0573, "Yucca Mountain Repository License Application." Rev. 1. ML090700817. Las Vegas, Nevada: DOE, Office of Civilian Radioactive Waste Management.
[3] NRC. 2003. NUREG–1804, "Yucca Mountain Review Plan—Final Report." Rev. 2. Washington, DC: NRC.

describe the major steps for completing each significant work element. DOE proposed a phased construction schedule with four phases. An initial operating capability will be developed during phase one, and full operating capabilities will be developed in phases two through four. DOE also provided a high-level project schedule of plans for construction, waste receipt, and emplacement operations, up to year 2030. During the first phase of construction, DOE intends to update the license application to request a license to receive and possess source, special nuclear or byproduct material, so it can begin receiving, handling, and emplacing waste packages when the first phase of construction is complete. DOE expects waste handling and emplacement operations to continue for 50 years. The NRC staff recognizes that DOE's emplacement schedules will depend on managing the heat load from spent fuel of differing ages and other factors that will be better known as operations proceed. The NRC staff considers DOE's emplacement schedule sufficient, given the information that is reasonably available now. For these reasons, the NRC staff finds that DOE provided, with respect to a construction authorization, schedules for construction, receipt of waste, and waste emplacement at the GROA that are sufficiently detailed to allow the NRC staff to evaluate the overall construction program for the GROA and its infrastructure.

Description of Security Measures for Physical Protection of Waste

The NRC staff reviewed DOE's description of planned security measures for the physical protection of spent nuclear fuel (SNF) and HLW at the Yucca Mountain repository. The intent of these measures is to ensure that theft or damage to SNF and HLW does not occur at the proposed repository. Plans must include the design for physical protection; contingency plans for specified actions in the event of threats, theft, or sabotage; and the security organization personnel training and qualification plan. The plan must list tests, inspections, audits, and other means to be used to demonstrate compliance with such requirements.

In addition to describing planned security measures consistent with regulatory requirements, DOE committed to provide a Physical Protection Plan, compliant with applicable portions of NRC's 10 CFR Part 73 regulations, after issuance of a construction authorization. On the basis of its review of DOE's descriptions of planned security measures and DOE's commitments, documented in the Appendix of this SER volume, the NRC staff finds that DOE's description of its planned security measures for physical protection of SNF and HLW at the Yucca Mountain repository is reasonably complete and therefore acceptable with respect to a construction authorization.

Description of the Material Control and Accounting Program

The NRC staff reviewed DOE's description of how it will maintain control of waste materials at the proposed repository and how it will document that all SNF and HLW are properly accounted for. NRC requires that DOE implement a program of material control and accounting (and accidental criticality reporting) that is the same as that specified elsewhere in NRC regulations at 10 CFR Parts 72 and 74. The NRC staff reviewed DOE's description to assure that (i) material balance, inventory, and record-keeping procedures for SNF and HLW are adequate and (ii) sufficient procedures are described for timely and adequate reporting of material status, material transfers and any accidental criticality or loss of special nuclear material.

In addition to describing its program consistent with regulatory requirements, DOE committed to provide its Material Control and Accounting Program, compliant with applicable portions of 10 CFR Part 74, after issuance of a construction authorization. On the basis of the NRC staff's review of DOE's description of its Material Control and Accounting Program, and DOE's commitments, as documented in the Appendix of this SER volume, the NRC staff finds that DOE's description of its Material Control and Accounting Program is reasonably complete and therefore acceptable with respect to a construction authorization.

Description of Work Conducted To Characterize the Yucca Mountain Site

The NRC staff evaluated DOE's description of work conducted to characterize the Yucca Mountain site. On the basis of this evaluation, the NRC staff determined that DOE provided an adequate description of work performed at Yucca Mountain to characterize the following aspects of the site: (i) geology; (ii) hydrology; (iii) geochemistry; (iv) geotechnical properties and conditions of the host rock; (v) climatology, meteorology, and other environmental sciences; and (vi) reference biosphere. The reference biosphere comprises the set of characteristics of the environment including, but not limited to, climate; topography; soils; flora; fauna; and human activities, such as diet and living style, similar to those experienced by people now living in Amargosa Valley, Nevada. The NRC staff finds that DOE's description of work conducted to characterize the Yucca Mountain site and the summary of the results from that work are adequate and acceptable with respect to a construction authorization.

Conclusion

On the basis of its review of the general information DOE provided in its June 3, 2008, license application and as updated on February 19, 2009, and on the basis of commitments specified in the Appendix, the NRC staff has made the following finding.

DOE has adequately described the proposed geologic repository at Yucca Mountain as specified in 10 CFR 63.21(b) of NRC's regulations because it has included:

1. A general description of the proposed geologic repository at the Yucca Mountain site, identifying the location of the geologic repository operations area, the general character of the proposed activities, and the basis for the exercise of the Commission's licensing authority.

2. Proposed schedules for construction, receipt of waste, and emplacement of wastes at the proposed geologic repository operations area.

3. A description of the detailed security measures for physical protection of high-level radioactive waste in accordance with 10 CFR 73.51 and generally described the design for physical protection, the safeguards contingency plan, the security organization personnel training and qualification plan, how the physical protection system is performance-tested to provide assurance that the system functions as intended, and how the system is tested and maintained to ensure its continued effectiveness, reliability, and availability.

4. A description of the material control and accounting program to meet the requirements of 10 CFR 63.78.

5. A description of work conducted to characterize the Yucca Mountain site.

ACRONYMS AND ABBREVIATIONS

CRCF	Canister Receipt and Closure Facility
DOE	U.S. Department of Energy
EIS	Environmental Impact Statement
FEPs	features, events, and processes
GI	General Information
GROA	geologic repository operations area
HLW	high-level radioactive waste
MTHM	metric tons of heavy metal
NMMSS	Nuclear Materials Management and Safeguards System
NRC	U.S. Nuclear Regulatory Commission
RAI	request for additional information
RMEI	reasonably maximally exposed individual
SAR	Safety Analysis Report
SER	Safety Evaluation Report
SNF	spent nuclear fuel
YMRP	Yucca Mountain Review Plan

INTRODUCTION

On June 3, 2008, the U.S. Department of Energy (DOE) submitted a license application to the U.S. Nuclear Regulatory Commission (NRC) seeking an authorization to begin construction of a geologic repository for high-level radioactive waste (HLW) disposal at Yucca Mountain, Nevada.[1] On February 19, 2009, DOE submitted its first update to the application.[2] In accord with requirements in 10 CFR Part 63, "Disposal of High-Level Radioactive Wastes in a Geologic Repository at Yucca Mountain, Nevada," the license application consists of general information and a Safety Analysis Report (SAR).

Disposal of HLW in a geologic repository at Yucca Mountain, Nevada, is governed by the rules in 10 CFR Part 63. NRC's regulation at 10 CFR Part 63 prescribes the requirements governing the licensing (including issuance of a construction authorization) of the DOE to receive and possess source, special nuclear, and byproduct material at a geologic repository operations area sited, constructed, or operated at Yucca Mountain, Nevada. According to 10 CFR Part 63 there are several stages in the licensing process. The site characterization stage, when the performance confirmation program is started, begins before submission of a license application. The construction stage would follow after the issuance of a construction authorization. A period of operations follows the Commission's issuance of a license to receive and possess source, special nuclear, and byproduct material. The period of operations includes the time during which emplacement of wastes occurs; any subsequent period before permanent closure during which the emplaced wastes are retrievable; and permanent closure, which includes sealing openings to the repository. Permanent closure represents the end of the performance confirmation program; final backfilling of the underground facility, if appropriate; and the sealing of shafts, ramps, and boreholes and follows the Commission's issuance of a license amendment for permanent closure.

In summary, the multi-staged licensing approach comprises four major decisions by the Commission: (i) construction authorization; (ii) license to receive and emplace waste; (iii) license amendment for permanent closure; and (iv) termination of the license. The multi-staged licensing process affords the Commission the flexibility to make decisions in a logical time sequence that accounts for DOE collecting and analyzing additional information over the construction and operational phases of the repository. At each stage, DOE must provide sufficient information to support that stage. Thus, as described at 10 CFR 63.21(a), the application must be as complete as possible in the light of information that is reasonably available at the time of docketing.

The NRC staff documents its review and evaluation of a license application in a Safety Evaluation Report (SER). This SER evaluates the DOE's request for a construction authorization pursuant to 10 CFR 63.31. As noted in a July 2009 Atomic Safety Licensing Board order,[3] the NRC staff plans to issue its SER in five volumes. SER Volume 1 presents the results of the NRC staff's review of the general information DOE provided in its license application. In SER Volumes 2–5, the NRC staff plans to present the results of its safety review of the DOE SAR. Any NRC decision on whether to authorize construction of a geologic repository for high-level radioactive waste (HLW) disposal at Yucca Mountain, Nevada, will be

---

[1] DOE. 2008. DOE/RW–0573, "Yucca Mountain Repository License Application." Rev. 0. ML081560400. Las Vegas, Nevada: DOE, Office of Civilian Radioactive Waste Management.
[2] DOE. 2009. DOE/RW–0573, "Yucca Mountain Repository License Application." Rev. 1. ML090700817. Las Vegas, Nevada: DOE, Office of Civilian Radioactive Waste Management.
[3] NRC. 2009. "July 21, 2009 Board Order Concerning Serial Case Management." ML092020323. Washington, DC: NRC.

made only after the NRC staff has completed all volumes of the SER. A decision to issue a construction authorization will not be effective until after the Commission has completed its review under 10 CFR 2.1023. In conducting its review, the NRC staff was guided by the review methods and acceptance criteria contained in the Yucca Mountain Review Plan.[4] When requested by the NRC staff, DOE provided additional information to clarify or supplement the license application.

The general information in the license application provides an overview of DOE's engineering design concept for the repository. In the general information, DOE described the aspects of the Yucca Mountain site and its environment relevant to repository design and performance. Understanding the performance of the proposed repository, as designed, in the context of the Yucca Mountain site and its environment, allowed DOE to make risk-informed, performance-based judgments regarding compliance with the regulations. The NRC staff, in SER Volumes 2–5, subsequently plans to evaluate these judgments.

There are five requirements in 10 CFR 63.21(b) that specify what DOE must provide as part of the general information in its license application:

1. A general description of the proposed geologic repository at the Yucca Mountain site [10 CFR 63.21(b)(1)]

2. Proposed schedules for construction, receipt of waste, and emplacement of wastes [10 CFR 63.21(b)(2)]

3. A description of the detailed security measures for physical protection of HLW [10 CFR 63.21(b)(3)]

4. A description of the material control and accounting program [10 CFR 63.21(b)(4)]

5. A description of work conducted to characterize the Yucca Mountain site [10 CFR 63.21(b)(5)]

Accordingly, the general information material the NRC staff reviewed is largely informational in nature, with the more detailed technical discussions and descriptions found elsewhere in the SAR. The subsequent chapters in SER Volume 1 document the results of the NRC staff's review of the general information material in the DOE license application. The NRC staff also reviewed information DOE provided in response to NRC staff's requests for additional information. As appropriate, the SER provides specific citations to these additional information sources in the context of the NRC staff's review.

Chapter 1 evaluates the DOE's information on general description. Chapter 2 evaluates the DOE's information on proposed schedules for construction, receipt, and emplacement of waste. Chapter 3 evaluates the DOE's information on physical protection. Chapter 4 evaluates the DOE's information on material control and accounting. Chapter 5 evaluates the DOE's information on site characterization. Chapter 6 presents the NRC's staff's conclusions derived from its review of the applicant's general information. Chapter 7 provides a glossary of some terms used in the SER. The Appendix to SER Volume 1 provides DOE's commitments made on general information related to the construction and operation of the geologic repository at Yucca Mountain.

---

[4]NRC. 2003. NUREG–1804, "Yucca Mountain Review Plan—Final Report." Rev. 2. Washington, DC: NRC.

CHAPTER 1

1.1 General Description

1.1.1 Introduction

This chapter of the Safety Evaluation Report (SER) evaluates the U.S. Department of Energy's (DOE or applicant) information on general description. The U.S. Nuclear Regulatory Commission (NRC) staff's evaluation is based on information provided in General Information (GI) Section 1 (DOE, 2009av), as supplemented by the DOE response to a NRC staff request for additional information (DOE, 2009au). GI Section 1 contains the applicant's general description of the proposed geologic repository at the Yucca Mountain site, providing a general description of the geologic repository operations area (GROA) and its location, the general nature of the proposed activities to be performed at the GROA, and the basis for the exercise of the NRC's licensing authority over a repository.

The intent of providing general information in the license application is twofold. First, it allows the applicant to provide an overview of its engineering design concept for the repository, which NRC staff reviews in this chapter. Second, it allows the applicant to demonstrate its understanding of what aspects of the Yucca Mountain site and its environs influence repository design and performance, which NRC staff reviews in SER Section 1.5, "Description of Site Characterization Work."

Understanding the performance of the design, in the context of the Yucca Mountain site and its environs, allows the applicant to make risk-informed, performance-based judgments regarding compliance with the regulations in its Safety Analysis Report (SAR). The NRC staff evaluates the SAR in SER Volumes 2 to 5. Accordingly, the applicant's information in GI Section 1 that the NRC staff reviews in this chapter is generally informational in nature, with the more detailed technical discussions and descriptions found elsewhere in the SAR.

1.1.2 Regulatory Requirements

The requirements for general description are in 10 CFR 63.21(b)(1), which states that the general information must include a general description of the proposed geologic repository at the Yucca Mountain site, identifying the location of the GROA, the general character of the proposed activities, and the basis for the exercise of the Commission's licensing authority.

The NRC staff has followed the review guidance provided in the Yucca Mountain Review Plan (YMRP) (NRC, 2003aa). As described in YMRP Section 1.1.1, because the material to be reviewed is informational in nature no detailed technical analysis of the information addressed in YMRP Section 1.1 is required. YMRP Section 1.1.3 identifies the following three criteria that the NRC staff may consider in its evaluation:

1. The location and arrangement of the GROA are adequately defined.

2. The general nature of the activities to be conducted at the geologic repository is adequately described.

3. An adequate basis for the exercise of the NRC licensing authority is provided.

1.1.3 Technical Review

In SER Section 1.1.3.1 the NRC staff summarizes the applicant's information in GI Section 1 prior to documenting its review relative to the YMRP Section 1.1.3 acceptance criteria. In SER Section 1.1.3.2 the NRC staff summarizes its review methodology. The NRC staff presents its evaluation in subsequent separate sections (SER Sections 1.1.3.2.1–1.1.3.2.3) corresponding to the individual topics of the YMRP Section 1.1.3 acceptance criteria. In SER Section 1.1.3.2.4 the NRC staff evaluates the applicant's information in GI Section 1.4.1.

1.1.3.1 Summary of the DOE License Application on General Description

The applicant provided a general description of the GROA and its location in GI Section 1.1. In GI Section 1.2 the applicant described the general nature of the activities to be conducted at the GROA. The applicant described the basis for the exercise of the NRC licensing authority in GI Section 1.3. Finally, in GI Section 1.4.1 the applicant described general references.

1.1.3.2 NRC Staff Evaluation of General Description

In the following sections the NRC staff documents its evaluation of the applicant's general description. In its review in SER Section 1.1.3.2.1, the NRC staff confirms whether the applicant's general description of the location and arrangement of the GROA addressed the six subcriteria of YMRP Section 1.1.3, Acceptance Criterion 1. Next in SER Section 1.1.3.2.2, the NRC staff reviews the applicant's summary of the general nature of the activities to be conducted at the GROA and confirms whether the applicant's summary addressed the seven subcriteria of YMRP Section 1.1.3, Acceptance Criterion 2. Then in SER Section 1.1.3.2.3, the NRC staff reviews the applicant's description of the basis for the NRC's licensing authority relative to YMRP Section 1.1.3, Acceptance Criterion 3. In conducting its review relative to YMRP Section 1.1 guidance, the NRC staff confirms, to the extent that the applicant cited other sections of GI within GI Section 1, that the information is consistent and appropriate.

Although the YMRP does not identify criteria to evaluate the applicant's information in GI Section 1.4.1, the NRC staff evaluates the information for factual accuracy.

1.1.3.2.1 Location and Arrangement of the GROA

In GI Section 1.1 the applicant provided information on the location and arrangement of structures, systems, and components of the GROA. The applicant presented a general discussion of the physical characteristics of the repository site and environs in GI Section 1.1.1. In GI Section 1.1.2 the applicant provided a general description of structures, systems, and components of the surface facilities in the GROA and generally described the subsurface facilities in the GROA in GI Section 1.1.3. The applicant identified, in GI Sections 1.1.2.2 and 1.1.3.2, surface facilities and subsurface facilities, respectively, to be dismantled prior to decommissioning and closure. As part of the description of the surface GROA facilities and subsurface facilities in the GROA, the applicant defined the purpose of each GROA structure, system, and component and any interrelationships among them. In GI Section 1.1.4 the applicant provided a general discussion of its plans to restrict access to and regulate land uses around the GROA. The applicant provided general information on its radiological monitoring activities in GI Section 1.1.5.1 and summarized, in GI Section 1.1.5.2, its plans for mitigation of radiological emergency events.

The NRC staff's evaluation is based on (i) confirming whether the applicant's general description of the location and arrangement of the GROA addressed the subcriteria for YMRP Section 1.1.3, Acceptance Criterion 1, and (ii) determining whether the applicant's general description is accurate. In the following paragraphs, the NRC staff evaluates the applicant's information against each of the subcriteria. The NRC staff assesses the applicant's description of the location of the GROA, reviews the applicant's information relative to the physical characteristics of the site, and then completes its evaluation in the order of the remaining subcriteria for YMRP Section 1.1.3, Acceptance Criterion 1.

The NRC staff confirms that the applicant provided maps in GI Figures 1-2 and 1-4 that identified Federal land immediately surrounding Yucca Mountain and depicted the location of the GROA and controlled area boundaries. The applicant described the Yucca Mountain site as that area surrounding the GROA for which DOE exercises authority over its use in accordance with 10 CFR Part 63. Initially, the NRC staff identified that GI Figures 1-2 and 1-4 were inaccurate with regard to Federal ownership, the site boundary, and location of the controlled areas; however, the applicant has committed (DOE, 2009au) to update the license application to reflect the private ownership and the correct acreage of Patent 27-83-0002 in GI Figures 1-2 and 1-4 and revise the figures to show that the Patent 27-83-0002 area is private land excluded from the proposed land withdrawal area. The applicant's commitment is listed in SER Volume 1, Appendix. On the basis of the applicant's commitment (DOE, 2009au) to revise GI Figures 1-2 and 1-4 to accurately reflect ownership of the land, site boundary, and the location of controlled areas, the NRC staff finds the applicant has provided accurate information showing the location of the site and general location of the GROA. The NRC staff confirms that the applicant presented in GI Figure 1-5 a map detailing the location of GROA, and in GI Figure 1-6 a drawing of the surface facilities of the GROA. Thus, the NRC staff finds that the applicant provided an accurate description, using scaled drawings or maps, showing the location of the GROA and its associated structures, systems, and components.

The NRC staff confirms that the applicant provided a general discussion of the physical characteristics of the site and the natural setting. The NRC staff finds that the applicant described (i) the site location and geography in GI Section 1.1.1.1, (ii) site geology and hydrology in GI Section 1.1.1.2, and (iii) meteorology and climatology in GI Section 1.1.1.3. For instance, the NRC staff confirms that the applicant described in GI Section 1.1.1.2 where the repository is located within Yucca Mountain (i.e., in the unsaturated zone above the water table) and the types of rock (i.e., volcanic rocks) which comprise Yucca Mountain. In GI Section 1.1.1.2 the applicant stated that additional details concerning the geochemical characteristics of the Yucca Mountain Site are provided in GI Section 5.2.3, which the NRC staff reviews in SER Volume 1, Chapter 5. On the basis of the NRC staff's review of the information in GI Sections 1.1.1.1–1.1.1.3, the NRC staff finds that the applicant has provided a general discussion of the physical characteristics of the site and natural setting.

The NRC staff confirms that the applicant provided a summary of design features of the aboveground and belowground structures, systems, and components, which also identified whether the design features are permanent or temporary. The NRC staff finds that the applicant, in GI Sections 1.1.2.1–1.1.2.2, (i) generally described design features of the aboveground GROA (e.g., in Figure1-5 the applicant identified that underground ventilation shaft surface facilities are part of the surface GROA) and (ii) identified whether the design features are permanent or temporary. For instance, the NRC staff confirms that the applicant described, in GI Section 1.1.2.1, major surface design features, including the Initial Handling Facility, Aging Facilities, Receipt Facility, Canister Receipt and Closure Facilities (CRCFs), and miscellaneous repository facilities. The NRC staff confirms that the applicant described in GI Section 1.1.2.2 its process for determining which surface facilities would be no longer required and identified that, at permanent closure, a network of permanent monuments and markers will be erected. On the basis of the NRC staff's review of the information in GI Sections 1.1.2.1–1.1.2.2, the NRC staff finds that the applicant has provided a summary of the design features of the aboveground GROA and identified whether the design features are permanent or temporary.

The NRC staff also finds that the applicant, in GI Sections 1.1.3.1–1.1.3.2, (i) generally described design features of the belowground GROA and (ii) identified subsurface facilities to be dismantled prior to closure. For instance, the NRC staff confirms that the applicant described, in GI Section 1.1.3.1 major subsurface design features, including the emplacement drifts; nonemplacement excavations, including a performance confirmation observation drift and ventilation shafts; waste packages; and drip shields. The NRC staff confirms that the applicant described in GI Section 1.1.3.2 the closure process including installation of drip shields and removal of noncommitted materials from the subsurface facility. On the basis of its review of the information in GI Sections 1.1.3.1–1.1.3.2, the NRC staff finds that the applicant has provided a general discussion of the design features of the belowground GROA and identified whether the design features are permanent or temporary.

The NRC staff confirms that the applicant provided a definition of the purpose of each GROA structure, system, and component and any interrelationships among them. For instance, the NRC staff finds that the applicant, in GI Sections 1.1.2.1 and 1.1.3.1, generally described (i) the purpose of each GROA structure, system, and component and (ii) any interrelationships among them. For example, the NRC staff confirms that the applicant described for surface GROA facilities, in GI Section 1.1.2.1, that one purpose of the Receipt Facility is to provide safe and controlled operating areas for receiving and transferring rail-based transportation casks containing transportation, aging, and disposal canisters or dual-purpose canisters, and one purpose of the CRCF is to provide the space, radiological confinement, structures, and internal systems that support stand-alone canister handling operations. The applicant described that these two facilities are interrelated because the CRCF has the capability to receive dual-purpose and transportation, aging, and disposal canisters from the Receipt Facility. The NRC staff also confirms that the applicant described for the subsurface design features of the GROA, in GI Section 1.1.3.1, that the purpose of the emplacement drifts is for emplacement of 70,000 metric tons of heavy metal (MTHM) of waste contained in 11,000 waste packages, and that the purpose of ventilation shafts is to intake and exhaust air to meet thermal management goals for emplaced waste. On the basis of the NRC staff's review of the applicant's information in GI Sections 1.1.2.1 and 1.1.3.1, the NRC staff finds that the applicant provided a definition of the purpose of each GROA structure, system, and component and any interrelationships among them.

The NRC staff confirms that the applicant generally described its plans to restrict access to and regulate land uses around the GROA. The NRC staff finds that the applicant, in GI Sections 1.1.4, summarized (i) the Federal controlled land needed for the repository, (ii) the relationships between the GROA and the controls to restrict access to the GROA, and (iii) the network of permanent monuments and markers that will be erected to warn future generations of the presence and nature of the buried waste. The NRC staff assesses the accuracy of the applicant's statement that the area needed for the repository encompasses land controlled by three Federal agencies. The NRC staff initially identified that this statement was inaccurate with regard to Federal ownership, the site boundary, and location of the controlled areas; however, the applicant has committed (DOE, 2009au) to update the license application to reflect the private ownership and the correct acreage of Patent 27-83-0002 in GI Figures 1-2 and 1-4 and revise the figures to show that the Patent 27-83-0002 area is private land excluded from the proposed land withdrawal area. On the basis of the NRC staff's review of GI Section 1.1.4 and the applicant's commitment (DOE, 2009au) to revise GI Figures 1-2 and 1-4 to accurately reflect ownership of the land, site boundary, and the location of controlled areas, the NRC staff finds the applicant provided an accurate description of its plans to restrict access to and regulate land uses around the GROA. The applicant's commitment is listed in SER Volume 1, Appendix.

The NRC staff confirms that in GI Section 1.1.5 the applicant provided a general description of radiological monitoring instrumentation and activities, including an overview of how the applicant plans to address potential radiological emergency events during the operational life of the repository. In GI Section 1.1.5.1 the applicant identified that radiological monitoring activities are described in SAR Section 5.11 (DOE, 2008ab). The applicant identified in GI Section 1.1.5.2 that SAR Section 5.7 (DOE, 2009av) describes the Emergency Plan and addresses potential radiological emergency events during the operational life of the repository. The NRC staff's review of SAR Sections 5.11 and 5.7 will be documented in SER Sections 2.1.1.8 and 2.5.7, respectively. The NRC staff finds that the applicant, in GI Sections 1.1.5.1 and 1.1.5.2, generally described (i) radiological monitoring and (ii) plans for mitigation of radiological emergency events. For example, the NRC staff confirms that the applicant, as part of its general description of radiological monitoring, in GI Section 1.1.5.1 identified that the requirements for types of instruments necessary to support safe radiological operations and emergency response actions will be described in the Operational Radiation Protection Program (SAR Section 5.11). Also, the NRC staff confirms that the applicant, as part of its description of its plans of mitigation of radiological emergency events, identified in GI Section 1.1.5.2 means to mitigate consequences and identified that the Emergency Plan, described in SAR Section 5.7 (DOE, 2009av), will address potential radiological emergency events during the operational life of the repository. On the basis of the NRC staff's review of the applicant's general information in GI Sections 1.1.5.1 and 1.1.5.2, the NRC staff finds that, with respect to a construction authorization, the applicant provided a general description of radiological monitoring instrumentation and activities, including the applicant's overview of how it plans to address mitigation of radiological impacts associated with the construction and operation of the proposed repository.

The NRC staff confirms that the applicant's information on the general description of the GROA is consistent between the applicant's general information and DOE's Final Environmental Impact Statement (EIS). The NRC staff previously determined whether there were any differences between (i) the proposed action to be taken by the NRC and as described in the EISs and (ii) the proposed action described in the license application (NRC, 2008ac). The NRC staff's review (NRC, 2008ac) determined that there were no differences between the descriptions of the proposed actions in the EIS and in the general information. On the basis of that review, the NRC staff confirms that applicant's information in GI Section 1.1 is consistent with DOE's Final EIS.

Findings on Location and Arrangement of the GROA

On the basis of the NRC staff's review of the applicant's information in GI Section 1.1 and DOE (2009au), and consistent with YMRP Section 1.1.3, Acceptance Criterion 1, the NRC staff makes the following findings.

The NRC staff finds that the applicant provided an accurate general description of the GROA with respect to a construction authorization. The NRC staff finds that the applicant's description included: (i) a discussion of the physical characteristics of the site and natural setting; (ii) scaled drawings or maps showing the location of the GROA and its associated structures, systems, and components; (iii) a summary of the design features of the aboveground and belowground structures, systems, and components, with a designation of whether they are permanent or temporary; (iv) a definition of the purpose of each GROA structure, system, and component, and any interrelationships among them; (v) a summary of plans to restrict access to and regulate land uses around the GROA; and (vi) a general description of radiological monitoring instrumentation and activities, including the DOE's overview of how it plans to address mitigation of radiological impacts associated with the construction and operation of the proposed repository. Thus, the NRC staff concludes that the applicant adequately defined the location and arrangement of the GROA.

Evaluation Findings

The NRC staff reviewed DOE's GI Section 1.1 and other information submitted in support of the license application and finds with reasonable assurance that the requirements of 10 CFR 63.21(b)(1) are met with respect to a construction authorization. An adequate general description of the geologic repository has been provided that identifies the location of the GROA.

1.1.3.2.2   General Nature of the Activities To Be Conducted at the GROA

In GI Section 1.2 the applicant described the general nature of the GROA activities and identified that proposed schedules for construction, receipt, and emplacement of waste were provided in GI Section 2. The applicant presented a general description of the waste forms to be disposed in GI Section 1.2.1. In GI Section 1.2.2 the applicant provided a general description of routine operations and described its plans for the inspection and testing of waste forms and waste packages in GI Section 1.2.3. The applicant provided information on waste retrieval in GI Section 1.2.4. In GI Section 1.2.5 the applicant provided a description of its plans for repository closure. The applicant generally discussed the proposed use of the GROA for purposes other than disposal of spent nuclear fuel (SNF) and high-level radioactive waste (HLW) in GI Section 1.2.6. In GI Section 1.2.7 the applicant provided a description of its plans for emergency responses.

The NRC staff's evaluation focuses on confirming whether the applicant's summary description of the general nature of the activities to be conducted at the GROA addressed the seven subcriteria for YMRP Section 1.1.3, Acceptance Criterion 2. In the following paragraphs the NRC staff evaluates the applicant's information against each of these subcriteria.

The NRC staff confirms that the applicant provided a summary description of the types, kinds, and amounts of SNF and other HLW waste to be disposed. The NRC staff finds that the applicant summarily described in GI Section 1.2.1 (i) the types of SNF (i.e., commercial and DOE, including naval SNF); (ii) kinds of SNF (e.g., boiling water reactor assemblies and pressure water reactor assemblies); and (iii) amounts of SNF to be disposed (i.e., approximately 63,000 MTHM of commercial SNF and about 2,333 MTHM of DOE SNF). The NRC staff also confirms that the applicant described in GI Section 1.2.1 (i) the types of HLW (i.e., HLW from reprocessed commercial SNF and defense nuclear program HLW); (ii) kinds of HLW (e.g., vitrified glass); and (iii) amounts of HLW to be disposed (e.g., the DOE HLW allocation is 4,667 MTHM). On the basis of the NRC staff's review of the information in GI Sections 1.2.1, the NRC staff finds that the applicant has provided a summary description of the types, kinds, and amounts of SNF and other HLW waste to be disposed.

The NRC staff confirms that the applicant provided a summary description of the proposed operations that included receipt, handling, emplacement, and retrieval of waste and waste packages and that the description included basic plans for movement of personnel, material, and equipment during construction and normal operations. The NRC staff finds that the applicant summarily described in GI Section 1.2.2 the receipt, handling, potential aging, and emplacement of waste and waste packages. For instance, the NRC staff finds that the applicant summarized (i) how the mostly canistered waste stream is received (i.e., in transportation casks) in the GROA and is supported by different surface facilities (e.g., the Initial Handling Facility and CRCF) and (ii) how waste is emplaced underground (i.e., sealed waste packages are transferred to the transport emplacement vehicle in the surface GROA facilities, and the transport emplacement vehicle transports the waste package to the emplacement drifts within the mountain). The applicant described concurrent construction and operation, and the NRC staff confirms that the description included basic plans for movement of personnel, material, and equipment during construction and normal operations. For example, to ensure the safety of project personnel and operational security, the applicant stated that it will be necessary to separate construction and normal operations and that separation will be maintained by designing independent systems for repository operations and construction. The NRC staff finds that the applicant summarily described waste retrieval in GI Section 1.2.4. On the basis of the NRC staff's review of the information in GI Sections 1.2.2 and 1.2.4, the NRC staff finds that the applicant has provided a summary description of the proposed operations that included receipt, handling, emplacement, and retrieval of waste and waste packages and that the description included basic plans for movement of personnel, material, and equipment during construction and normal operations.

The NRC staff confirms that the applicant provided a description of plans for the inspection and testing of waste forms and waste packages. The NRC staff finds that the applicant summarily described in GI Section 1.2.3 the types of tests and inspections to be performed related to waste forms and the waste packages. For instance, the NRC staff confirms that the applicant stated that for the Initial Handling Facility and the CRCF, the applicant will verify the structural and surface conditions of canisters and waste packages, as well as determine the radiological levels of container content and the extent of potential surface contamination by conducting the following types of tests: (i) visual inspection and radiological surveys of the exteriors of incoming casks; (ii) sampling of cask internal gases (if required); (iii) nondestructive examination of waste package closure welds; and (iv) remote visual inspection of closed waste packages. On the basis of the NRC staff's review of the information in GI Section 1.2.3, the NRC staff finds that the applicant provided a description of plans for the inspection and testing of waste forms and waste packages.

The NRC staff confirms that the applicant provided a description of plans for the retrieval and the alternative storage of radioactive wastes, should retrieval be necessary. The NRC staff finds that in GI Section 1.2.4 the applicant stated that this section summarized the analysis provided in SAR Section 1.11 (DOE, 2008ab). The NRC staff confirms that the applicant's summary description included (i) how the GROA is designed to permit retrieval of any or all emplaced waste, starting at any time up to the beginning of permanent closure; (ii) reasons why retrieval operations could be initiated; and (iii) how, if a retrieval decision is made, waste would be placed in a storage or disposal facility designed in accordance with the regulations that are applicable at the time. On the basis of the NRC staff's review of the information in GI Sections 1.2.4, the NRC staff finds that the applicant provided a description of plans for the retrieval and the alternative storage of radioactive wastes, should retrieval be necessary.

The NRC staff confirms that the applicant provided a description of plans for decommissioning and permanent closure of the GROA. The NRC staff finds that in GI Section 1.2.5 the applicant summarily described repository closure. The NRC staff confirms that the applicant stated that its plans for permanent closure are described in SAR Section 1.12 (DOE, 2008ab). The NRC staff confirms that the applicant's summary description identified that (i) after repository operations and the performance confirmation program have been completed, DOE will file an application with NRC for a license amendment to close the repository; (ii) 10 CFR 63.51(a)(2)–(3) require that the applicant undertake measures to regulate or prevent activities that could impair long-term waste isolation and institute a monitoring program after permanent closure; and (iii) the applicant will erect a network of permanent markers, which were described in GI Section 1.1.4. On the basis of the NRC staff's review of the information in GI Section 1.2.5, the NRC staff finds that the applicant provided a description of plans for decommissioning and permanent closure of the GROA.

The NRC staff confirms that the applicant provided a general discussion of proposed uses of the GROA for purposes other than the disposal of SNF and other types of HLW. The NRC staff finds that in GI Section 1.2.6 the applicant generally discussed other uses of the GROA. For instance, the NRC staff confirms that the applicant stated it does not intend to use the GROA for purposes other than disposal of SNF and HLW, but identified that other uses of the GROA could include Native American cultural activities, independent performance monitoring by groups other than NRC and DOE, and activities related to the protection of flora and fauna. On the basis of the NRC staff's review of the information in GI Section 1.2.6, the NRC staff finds that the applicant has provided a general discussion of proposed uses of the GROA for purposes other than the disposal of SNF and other types of HLW.

The NRC staff confirms that in GI Section 1.2.7 the applicant provided information on plans for responses to emergencies. In GI Section 1.2.7 the applicant identified that a description of the Emergency Plan is provided in SAR Section 5.7. The NRC staff's review of SAR Section 5.7 will be documented in SER Section 2.5.7. The NRC staff confirms that the applicant stated in GI Section 1.2.7 that, as required by 10 CFR 63.161, the applicant is developing an Emergency Plan to respond to radiological accidents and that the plan is being developed in accordance with criteria contained in 10 CFR 72.32(b). The NRC staff confirms that the applicant identified in GI Section 1.2.7 the categories of information that will be included (e.g., a description of the types and classifications of potential accidents and a description of the means to mitigate consequences of each type of accident). On the basis of the NRC staff's review of the information in GI Section 1.2.7, the NRC staff finds that, with respect to a construction authorization, the applicant provided an overview of how it plans to respond to emergencies.

Findings on General Nature of the Activities To Be Conducted at the GROA

On the basis of the NRC staff's review of the applicant's information in GI Section 1.2, and consistent with YMRP Section 1.1.3, Acceptance Criterion 2, the NRC staff makes the following findings.

The NRC staff finds that the applicant provided a summary description of the types, kinds, and amounts of SNF and other HLW to be disposed. The NRC staff finds that the applicant provided a summary description of the proposed operations that includes receipt, handling, emplacement, and retrieval of waste and waste packages. In addition, the NRC staff finds that the applicant's description included basic plans for the movement of personnel, material, and equipment during construction and normal operations. The NRC staff finds that the applicant provided a description of plans for the inspection and testing of waste forms and waste packages. The NRC staff finds that the applicant provided a description of plans for the retrieval and the alternative storage of radioactive wastes, should retrieval be necessary. The NRC staff finds that the applicant provided a description of plans for decommissioning and permanent closure of the GROA. The NRC staff finds that the applicant incorporated a general discussion of proposed uses of the GROA for purposes other than the disposal of SNF and other types of HLW. The NRC staff finds that the applicant provided a summary of the description of plans for responses to emergencies. Thus, the NRC staff concludes that, with respect to a construction authorization, the applicant adequately described the general nature of the activities to be conducted at the GROA.

Evaluation Findings

The NRC staff reviewed DOE's GI Section 1.2 and other information submitted in support of the license application and finds with reasonable assurance that the requirements of 10 CFR 63.21(b)(1) are met with respect to a construction authorization. An adequate general description of the geologic repository has been provided that discusses the general character of the proposed activities at the GROA.

1.1.3.2.3   Basis for the Exercise of the NRC Licensing Authority

Consistent with the requirements of 10 CFR 63.21(b)(1), in GI Section 1.3, the applicant described the basis for NRC's exercise of its licensing authority. The NRC staff verifies that the applicant described (i) how Congress established the bases for NRC licensing of a repository at Yucca Mountain; (ii) Congress' passage of Public Law No. 107-200, 116 Stat. 735, which approved the site at Yucca Mountain, Nevada, as the candidate site for the location of a HLW repository; and (iii) the U.S. Environmental Protection Agency's standard (40 CFR Part 197); and (iv) NRC rule (10 CFR Part 63). The NRC staff also verifies that the applicant stated that the NRC determination that the DOE license application satisfies all of the requirements of 10 CFR Part 63, as well as all of the other NRC regulatory requirements applicable to the repository, constitutes an adequate basis for NRC exercise of its statutory licensing authority.

<u>Findings on Basis for the Exercise of the NRC Licensing Authority</u>

On the basis of the NRC staff's review of the information in GI Sections 1.3, and consistent with YMRP Section 1.1.3, Acceptance Criterion 3, the NRC staff finds that the applicant's license application described the basis for the Commission's licensing authority that applies to the proposed activities at the geologic repository. Thus, the NRC staff concludes that the applicant provided an adequate basis for the exercise of the NRC licensing authority.

Evaluation Findings

The NRC staff reviewed DOE's GI Section 1.3 and other information submitted in support of the license application and finds with reasonable assurance that the requirements of 10 CFR 63.21(b)(1) are met with respect to a construction authorization. An adequate general description of the geologic repository has been provided that includes the basis for the exercise of the Commission's licensing authority.

1.1.3.2.4   General References

In GI Section 1.4.1 the applicant stated general references are references that are not part of the license application and instead provide information or additional detail that will facilitate review of the license application. The applicant also identified what informational types of documents are general references, how the documents may be referenced (i.e., referenced in whole or in part), and where references to such information may be listed in the license application.

The NRC staff finds the applicant's description of general references acceptable because it is consistent with 10 CFR 63.21(a).

1.1.4 Evaluation Findings

The NRC staff reviewed DOE's GI Section 1.1 and other information submitted in support of the license application and finds with reasonable assurance that the requirements of 10 CFR 63.21(b)(1) are met with respect to a construction authorization. An adequate general description of the geologic repository has been provided that identifies the location of the GROA, discusses the general character of the proposed activities at the GROA, and provides the basis for the exercise of the Commission's licensing authority.

1.1.5 References

DOE. 2009au. "Yucca Mountain—Response to Request for Additional Information Regarding License Application (Safety Analysis Report Section 5.8), Safety Evaluation Report Vol. 4, Chapter 2.5.8." Letter (May 6) J.R. Williams to F. Jacobs (NRC). ML091330698. Washington, DC: DOE, Office of Technical Management.

DOE. 2009av. DOE/RW–0573, "Yucca Mountain Repository License Application." Rev. 1. ML090700817. Las Vegas, Nevada: DOE, Office of Civilian Radioactive Waste Management.

DOE. 2008ab. DOE/RW–0573, "Yucca Mountain Repository License Application." Rev. 0. ML081560400. Las Vegas, Nevada: DOE, Office of Civilian Radioactive Waste Management.

NRC. 2008ac. "Staff's Adoption Determination Report for the U.S. Department of Energy's Environmental Impact Statements for the Proposed Geologic Repository at Yucca Mountain." ML082420342. Washington, DC: NRC.

NRC. 2003aa. NUREG–1804, "Yucca Mountain Review Plan—Final Report." Rev. 2. Washington, DC: NRC.

CHAPTER 2

1.2 Proposed Schedules for Construction, Receipt, and Emplacement of Waste

1.2.1 Introduction

This chapter of the Safety Evaluation Report (SER) evaluates the U.S. Department of Energy's (DOE or applicant) information on proposed schedules for construction, receipt, and emplacement of waste. The U.S. Nuclear Regulatory Commission (NRC) staff's evaluation is based on information provided in General Information (GI) Section 2 (DOE, 2009av). GI Section 2 contains the applicant's description of its proposed schedules for site preparation; construction of structures, systems, and components of the geologic repository operations area (GROA) (including development of requisite infrastructure both on- and off-site of the GROA); and its proposed schedules for the receipt, handling, and emplacement of waste package canisters.

This chapter of the SER, along with SER Sections 1.1, "General Description," and 1.5, "Description of Site Characterization Work," addresses the applicant's overview of its engineering design concept for the repository and its understanding of what aspects of the Yucca Mountain site and its environs influence repository design and performance.

1.2.2 Regulatory Requirements

The requirements for proposed schedules are in 10 CFR 63.21(b)(2), which states that the license application must include proposed schedules for construction, receipt of waste, and emplacement of wastes at the proposed geologic repository operations area.

The NRC staff reviewed the applicant's proposed schedules using guidance in the Yucca Mountain Review Plan (YMRP) Section 1.2 (NRC, 2003aa). As described in YMRP Section 1.2.1, because the material to be reviewed is informational in nature no detailed technical analysis of the information addressed in YMRP Section 1.2 is required. YMRP Section 1.2.3 identifies one criterion that the NRC staff may consider in its evaluation: major steps for the completion of each significant work element are adequately described.

1.2.3 Technical Review

This section presents the NRC staff's evaluation of information the applicant presented on proposed schedules for construction, receipt, and emplacement of waste.

1.2.3.1 Summary of the DOE License Application on Proposed Schedules

The applicant proposed a phased construction schedule with four phases. The applicant's approach will allow for an initial operating capability to be developed during phase one, while full operating capabilities will be developed in phases two through four. The phased schedule will allow the facility to begin receiving, packaging, and emplacing a limited throughput of spent nuclear fuel and high-level waste while construction associated with phases two through four is being completed. GI Section 2 Figure 2-2 (DOE, 2009av) showed the proposed layout of the surface facility and the facilities proposed to be built during each phase. The proposed layout and construction phasing of the subsurface facilities were shown in GI Section 2 Figure 2-3 (DOE, 2009av).

In GI Figure 2-1 the applicant provided a high-level project schedule of plans for construction, waste receipt, and emplacement operations, up to year 2030. For instance, Figure 2-1 projects that NRC issues a construction authorization in 2011, and in 2016, the applicant submits its license application to NRC to receive and possess source, special nuclear material, or byproduct material at the site. According to its proposed schedule, DOE intends to update the license application to request a license to receive and possess source, special nuclear material, or byproduct material at the site during the first phase of construction, so that it can begin receiving, handling, and emplacing waste packages when the first phase of construction is complete. The applicant expects waste handling and emplacement operations to continue for 50 years.

In GI Section 2.1.1 "Initial Operating Capability," the applicant described phase one construction. During the first phase the applicant plans to construct the Initial Handling Facility, one Canister Receipt and Closure Facility (CRCF), the Wet Handling Facility, part of the Aging Facility, and components of subsurface Panel 1, along with infrastructure improvements and support facilities. DOE further stated that programs for physical protection (evaluated by the NRC staff in SER Section 1.3), material control and accounting (evaluated by the NRC staff in SER Section 1.4), and emergency planning (evaluated by the NRC staff in SER Section 2.5.7) will be in place during this initial phase, before operations to receive and possess waste begin. The applicant described the facilities and infrastructure that would be constructed in the subsequent construction phases in GI Section 2.1.2 "Full Operating Capability."

1.2.3.2   NRC Staff Evaluation of Proposed Schedules

The NRC staff reviewed the applicant's license application using the review method and acceptance criterion in YMRP Sections 1.2.2 and 1.2.3. As described in YMRP Sections 1.2.1 and Section 1.2.2 the NRC staff recognizes that schedules will evolve and become more detailed over time and the NRC staff recognizes that scheduling will be a function of evolving circumstances and expects distant scheduling to be less detailed than near-term scheduling.

The NRC staff verifies that the applicant provided, in GI Figures 2-1 to 2-3, a schedule and time-scaled charts of planned construction activities, which include major steps during construction. For instance, the NRC staff finds that in Figure 2-1 the applicant provided the schedule for construction of the Initial Handling Facility and its major steps (design, procurement, construction, and startup). The NRC staff also finds that in Figure 2-1 the applicant provided the major steps for the completion of each of the other significant work elements (e.g., CRCF and Wet Handling Facility).

The NRC staff verifies that the information supplied (specifically, the scheduled time allocated for each work step) is sufficient for an overall understanding of the GROA and infrastructure construction and emplacement operations. For instance, the NRC staff determines that in Figure 2-1, in conjunction with Figure 2-3, the applicant provided the major steps and activities, along with their durations, associated with the receipt and emplacement of wastes. The NRC staff recognizes that specific emplacement schedules are highly dependent on thermal management and other factors that will be better known as operations proceed. The NRC staff finds the emplacement schedule provided to be sufficient with respect to a construction authorization.

The applicant, in GI Sections 2 and 2.1, described the phased construction approach and its relationship to development of the initial, and full, operating capabilities of the repository and provided a schedule in Figure 2-1 for all the phases of construction. The NRC staff uses that information to verify that construction of GROA facilities will be substantially complete before the proposed scheduled receipt and emplacement of wastes. For instance, the NRC staff verifies that in Figure 2-1, for the initial operating capability, drifts 1-3 (of a total of four drifts) of Panel 1 would be completed before the scheduled dates of receipt and emplacement of wastes.

The NRC staff concludes that the applicant has provided and described the major steps for construction, and for the receipt and emplacement of wastes. The applicant provided an integrated high-level project schedule and time-scaled charts of planned construction activities. From these and the description of the phased construction approach in GI Sections 2 and 2.1, the NRC staff concludes that the scheduled time allocated for each major activity and the identified interdependence of major activities are sufficient to provide an overall understanding of the GROA and infrastructure construction and routine waste emplacement operations. Therefore, the NRC staff also concludes that, for each of the activities described in each phase of GROA operations and activities, the applicant adequately described planned overall project progress for each activity described in the various phases, in accord with the acceptance criterion of YMRP Section 1.2.3.

1.2.4 Evaluation Findings

The NRC staff has reviewed GI Section 2 and other information submitted in support of the license application and has found, with reasonable assurance, that the requirements of 10 CFR 63.21(b)(2) are satisfied with respect to a construction authorization. DOE provided schedules for construction, receipt of waste, and waste emplacement at the GROA that are sufficiently detailed to allow the NRC staff to evaluate the overall construction program for the GROA and its infrastructure.

1.2.5 References

DOE. 2009av. DOE/RW–0573, "Yucca Mountain Repository License Application." Rev. 1. ML090700817. Las Vegas, Nevada: DOE, Office of Civilian Radioactive Waste Management.

NRC. 2003aa. NUREG–1804, "Yucca Mountain Review Plan—Final Report." Rev. 2. Washington, DC: NRC.

CHAPTER 3

1.3 Physical Protection Plan

1.3.1 Introduction

This chapter of the Safety Evaluation Report (SER) evaluates the U.S. Department of Energy's (DOE or applicant) information on physical protection. The U.S. Nuclear Regulatory Commission (NRC) staff's evaluation is based on information provided in General Information (GI) Section 3 (DOE, 2008ab). GI Section 3 contains the applicant's description of the detailed security measures for physical protection of spent nuclear fuel (SNF) and high-level radioactive waste (HLW) at the Yucca Mountain repository. The applicant's information on physical protection of the repository at Yucca Mountain is part of the information that the NRC staff will consider, once the NRC staff's evaluation is completed for all volumes of the SER, to determine whether the common defense and security requirements, for a construction authorization, have been met.

SER Volume 1, Chapters 1–2, provide information that allows the NRC staff to determine, in this chapter, whether the applicant's information on physical protection is consistent with the applicant's overview of its engineering design concept and schedule for the repository.

1.3.2 Regulatory Requirements

The requirements for physical protection are in 10 CFR 63.21(b)(3), which states that the general information must include a description of the detailed security measures for physical protection of HLW in accordance with 10 CFR 73.51. This plan must include the design for physical protection, the licensee's safeguards contingency plan, and security organization personnel training and qualification plan. The plan must list tests, inspections, audits, and other means to be used to demonstrate compliance with such requirements.

10 CFR 73.51 specifies the requirements for the physical protection of stored SNF and HLW at a geologic repository operations area (GROA). The general performance objectives are specified at 10 CFR 73.51(b)(1) and 10 CFR 73.51(b)(3). 10 CFR 73.51(b)(1) specifies that each licensee shall establish and maintain a physical protection system with the objective of providing high assurance that activities involving SNF and HLW do not constitute an unreasonable risk to public health and safety. To meet this objective, the physical protection system must meet the following performance capabilities, in accordance with 10 CFR 73.51(b)(2):

- Store SNF and HLW only within a protected area

- Grant access to the protected area only to individuals who are authorized to enter the protected area

- Detect and assess unauthorized penetration of, or activities within, the protected area

- Provide timely communication to a designated response force whenever necessary

- Manage the physical protection organization in a manner that maintains its effectiveness

10 CFR 73.51(b)(3) states that the physical protection system must be designed to protect against loss of control of the facility that could be sufficient to cause a radiation exposure exceeding the dose as described in 10 CFR 72.106(b).

10 CFR 73.51(c) requires that the licensee retain a copy of the effective physical protection plan for 3 years or until termination of the license for which procedures were developed. 10 CFR 73.51(d)(1)–(13) lists the methods acceptable to NRC for meeting the performance capabilities of 10 CFR 73.51(b)(2).

The NRC staff reviewed the applicant's physical protection plan using guidance in the Yucca Mountain Review Plan (YMRP) Section 1.3 (NRC, 2003aa). As described in YMRP Section 1.3 the NRC staff's review is intended to determine with reasonable assurance whether the applicant has provided a description of the detailed security measures for physical protection that (i) is complete in light of information that is reasonably available at the time of docketing and (ii) provides assurance that activities involving HLW do not present an unreasonable risk to the public health and safety. YMRP Section 1.3.3 identifies 11 criteria that the NRC staff may consider in its evaluation:

1. The physical protection plan contains an adequate GROA description and provides an acceptable schedule for implementation.

2. General performance objectives will be met.

3. The protection goal will be met.

4. The security organization will be adequate.

5. Physical barrier subsystems will be adequate.

6. Access control subsystems and procedures will be adequate.

7. Detection, surveillance, and alarm subsystems and procedures will be adequate.

8. Communication subsystems will be adequate.

9. Equipment operability and compensatory measures are adequate.

10. Contingency and response plans and procedures will be adequate.

11. Reporting of safeguards events will be adequate.

As noted in YMRP Section 1.3 in light of the terrorist attacks of September 11, 2001, the Commission directed the NRC staff to conduct a comprehensive reevaluation of NRC physical protection requirements. On December 20, 2007, NRC published a proposed rule (NRC, 2007ae) to revise security requirements for the GROA. The proposed changes include specific training enhancements, improved access authorization, enhancements to defensive strategies, and enhanced reporting requirements, and affect portions of 10 CFR Parts 63 and 73 that apply to the Yucca Mountain repository. Because the proposed rule has not been finalized at the time of the NRC staff review of the license application, the applicant was not required to address the proposed changes. However, as described in SER Section 1.3.3.1, DOE has committed to submit a detailed Physical Protection Plan, compliant with applicable portions of 10 CFR Part 73, no later than 180 days after the issuance of a construction authorization. The NRC staff will review the adequacy of DOE's detailed Physical Protection Plan consistent with applicable requirements of 10 CFR Parts 63 and 73.

1.3.3 Technical Review

In SER Section 1.3.3.1 the NRC staff summarizes the applicant's overall approach for the Physical Protection Plan, and in SER Section 1.3.3.2 the NRC staff documents its review relative to the YMRP Section 1.3.3 acceptance criteria and summarizes its review methodology. The NRC staff's evaluation in subsequent separate sections (SER Sections 1.3.3.2.1–1.3.3.2.11) corresponds to the individual topics of the YMRP Section 1.3.3 acceptance criteria. In SER Section 1.3.3.2.12 the NRC staff summarizes its evaluation and presents its findings.

1.3.3.1 Summary of the DOE License Application on Physical Protection Plan

In GI Section 3, the applicant provided a description of the detailed security measures for physical protection of SNF and HLW at the Yucca Mountain repository. The applicant committed to submit to NRC a Physical Protection Plan, compliant with applicable portions of 10 CFR Part 73, no later than 180 days after issuance of a construction authorization. As identified in GI Figure 2-1, the applicant projects that NRC issues a construction authorization in 2011, and in 2016, the applicant submits its license application to NRC to receive and possess source, special nuclear material, or byproduct material at the site. The applicant identified that the Physical Protection Plan will describe the physical protection system for the GROA, which will be designed to protect against a loss of control of the facilities that could cause radiation exposures exceeding the doses described in 10 CFR 72.106. In describing its Physical Protection Plan, the applicant identified how protected area operations, where waste handling occurs, will be separated from construction activities. The applicant identified that the security requirements for GROA areas will differ depending on whether nuclear material is present. In GI Section 3, the applicant stated that those security requirements and the resulting levels of protection will be described in the Physical Protection Plan. The applicant stated that the attributes of the Physical Protection Plan described in GI Sections 3.1–3.12 provide an overview of how the general performance requirements, performance capabilities, and specific requirements in the regulations will be developed and implemented and that its description forms the basis for meeting the requirements of 10 CFR 63.21(b)(3) and 10 CFR 73.51.

1.3.3.2 NRC Staff Evaluation of Physical Protection Plan

In the following sections the NRC staff documents its evaluation of the applicants' description of its Physical Protection Plan against 10 CFR 63.21(b)(3). In conducting its review, the NRC staff compares the applicant's description to all the attributes of each criterion to ensure that the applicant's description of its Physical Protection Plan is complete and consistent with the guidance in YMRP Section 1.3.

1.3.3.2.1 Description and Schedule for Implementation

In GI Section 3.1 the applicant identified that the Physical Protection Plan will describe the GROA and described that the plan will be updated as the protected area within the GROA changes to accommodate the phased operations of the waste handling facilities. The applicant identified that the GROA layout will be described through a written description and use of maps and plot plans and that the types of facilities and physical protection systems (e.g., major components of GROA facilities and security posts and access control points), and their locations, will be included in the GROA description. The applicant stated that a schedule for implementation of physical protection will be included in the plan. In Section 3.1 the applicant stated that the security program will be operational and performance-tested prior to receipt of SNF and HLW. The applicant described how the Physical Protection Plan will be treated as a record and described its approach to record retention.

The NRC staff finds the applicant's information on the description of the GROA acceptable for two reasons. First, the NRC staff finds that the applicant identified in GI Section 3.1 that the GROA will be described in the Physical Protection Plan and will be updated as the protected area within the GROA changes to accommodate the phased operations of the waste handling facilities. The applicant's basis for updating the Physical Protection Plan in GI Section 3.1 is acceptable because it is consistent with the applicant's description of concurrent construction and operation in GI Section 1.2.2. Second, the NRC staff finds that the applicant, in GI Section 1.2.1, described the types of waste forms to be disposed of, and in GI Section 3.1, described the information that will be included (e.g., location of physical protection systems, subsystems, and major components of the GROA facilities) and how the information will be conveyed (e.g., maps and narrative). Thus, the NRC staff finds that the applicant has included an adequate description of the GROA that will be incorporated and updated in the Physical Protection Plan.

The NRC staff finds the applicant's schedule for implementation acceptable because the information the applicant will provide is consistent with the acceptance criterion on GROA description and implementation schedule. The following three examples demonstrate the applicant's consistency with the acceptance criterion on GROA description and implementation schedule. First, the applicant committed in GI Section 3 to submit to NRC a Physical Protection Plan, compliant with applicable portions of 10 CFR Part 73, no later than 180 days after issuance of a construction authorization; this will include an implementation schedule for the physical protection program. The applicant's commitment is listed in SER Volume 1, Appendix. The NRC staff finds the applicant's commitment acceptable for two reasons. The first reason is the NRC staff finds the applicant's commitment is consistent with the requirement of 10 CFR 63.21(b)(3), which requires DOE to provide a description of the detailed security measures to be used. 10 CFR Part 63 provides for a multistaged licensing process that affords the Commission the flexibility to make decisions in a logical time sequence that takes into account that DOE will collect and analyze additional information over the construction and operational phases of the repository, DOE has provided a sufficient description of its physical protection plan to support issuance of a construction authorization. The second reason is that the NRC staff finds that the Physical Protection Plan will be submitted, according to the applicant's schedule (GI Figure 2.1), before the design of any facilities associated with initial operating capability is completed and several years before the applicant will submit a license application amendment to receive and possess source, special nuclear material, or byproduct material. The second example is the NRC staff finds the applicant's statement that the security program will be operational and performance tested prior to receipt of SNF and HLW is consistent with the acceptance criterion and, thus, acceptable. The third example is the NRC staff finds the applicant's description of its schedule for implementation acceptable because the applicant stated that copies of the Physical Protection Plan and changes thereto will be treated as records for 3 years or until termination of the license, consistent with the requirement of 10 CFR 73.51(c). Thus, the NRC staff finds that the applicant's information on the schedule for implementation of the Physical Protection Plan acceptable.

Findings on Description and Schedule for Implementation

On the basis of the NRC staff's review of the applicant's information, and consistent with YMRP Section 1.3.3, Acceptance Criterion 1, the NRC staff makes the following finding.

On the basis of information provided in GI Sections 1.2.2, 3, and 3.1 and GI Figure 2.1, the applicant's commitment to submit to NRC a Physical Protection Plan, compliant with applicable portions of 10 CFR Part 73, no later than 180 days after issuance of a construction authorization, and its statement that the security program will be operational and performance-tested prior to receipt of SNF and HLW, the NRC staff finds that, with respect to a construction authorization, the Physical Protection Plan contains an adequate GROA description and provides an acceptable schedule for implementation.

Evaluation Findings

The NRC staff has reviewed the GI and other information submitted in support of the license application and finds, with reasonable assurance, that the requirements of 10 CFR 63.21(b)(3) are satisfied with respect to a construction authorization. DOE has described the detailed security measures for physical protection of HLW in accordance with 10 CFR 73.51(c), and DOE will retain a copy of the effective physical protection plan for 3 years or until termination of the license.

1.3.3.2.2 General Performance Objectives

In GI Section 3.2 the applicant described how the Physical Protection Plan addresses the general performance objectives and requirements of 10 CFR 73.51(b)(1)–(2).

The NRC staff finds that the applicant's description addressed how the physical protection system will provide assurance that activities involving SNF and HLW do not present an unreasonable risk to the public health and safety. For instance, in GI Section 3 the applicant described elements (e.g., effective implementation) that provide assurance that activities involving SNF and HLW do not present an unreasonable risk to the health and safety of the public. Because the applicant committed in GI Section 3 to submit to NRC a Physical Protection Plan, compliant with applicable portions of 10 CFR Part 73, no later than 180 days after issuance of a construction authorization, and the applicant stated in its description that the Physical Protection Plan will meet the general performance objectives and requirements in 10 CFR 73.51(b)(1) and will establish, implement, and maintain a physical protection system, which is consistent with YMRP Section 1.3.3, Acceptance Criterion 2, Subcriterion 2, the NRC staff finds the applicant's description of the general performance objectives acceptable.

The NRC staff finds that the applicant's description of the Physical Protection Plan stated the plan will describe those portions of the physical protection system for which redundant and diverse components and redundant and diverse subsystems and components are necessary to ensure adequate performance, as required by 10 CFR 73.51(b)(2). The NRC staff finds the applicant's description acceptable because it is consistent with YMRP Section 1.3.3, Acceptance Criterion 2, Subcriterion 3.

The NRC staff finds that the applicant's description of the Physical Protection Plan stated that the physical protection system will be designed and performance-tested to provide assurance that the system functions as intended. In addition the NRC staff identifies that the applicant stated the plan will describe the design and how the system is tested and maintained to ensure its continued effectiveness, reliability, and availability. These described attributes are consistent with YMRP Section 1.3.3, Acceptance Criterion 2, Subcriterion 4, and thus the NRC staff finds the applicant's description acceptable.

Findings on General Performance Objectives

On the basis of the NRC staff's review of the applicant's information and NRC staff's evaluation in SER Section 1.3.3.2.1 and consistent with YMRP Section 1.3.3, Acceptance Criterion 2, the NRC staff makes the following findings.

On the basis of information provided in GI Sections 3 and 3.2, the NRC staff finds that the general performance objectives will be met. The NRC staff finds that, with respect to a construction authorization, the applicant's description of its Physical Protection Plan acceptably addressed each of the subcriteria of YMRP Section 1.3.3, Acceptance Criterion 2.

Evaluation Findings

The NRC staff has reviewed the GI and other information submitted in support of the license application and finds, with reasonable assurance, that the requirements of 10 CFR 63.21(b)(3) are satisfied with respect to a construction authorization. DOE has described the detailed security measures for physical protection of HLW in accordance with 10 CFR 73.51(b)(1)–(2). DOE will establish and maintain a physical protection system with the objective of providing high assurance that activities involving SNF and HLW do not constitute an unreasonable risk to public health and safety and will meet the required performance capabilities.

1.3.3.2.3 Protection Goal

In GI Section 3.3 the applicant described how the Physical Protection Plan addresses the protection goal (i.e., the physical protection system must be designed to protect against loss of control of the facility that could be sufficient to cause a radiation exposure exceeding the dose as described in 10 CFR 72.106) and its strategy for meeting the goal.

The applicant's description of the Physical Protection Plan stated the implementation of the plan will address protection of the protected area of the GROA against a loss of control that could cause radiation exposure exceeding the dose defined in 10 CFR 72.106. As the applicant described, the plan will identify physical barriers; access controls; a security force; intrusion detection, assessment, and surveillance systems; communication equipment; and contingency and response plans and procedures to ensure that the protection goal is met. The applicant stated that the plan will describe the strategy for denying unauthorized access. The NRC staff finds the applicant's description of the protection goal within the Physical Protection Plan acceptable because each of the attributes the applicant described are the same as those identified in the protection goal acceptance criterion (YMRP Section 1.3.3, Acceptance Criterion 3) and because the applicant has committed to submit to NRC a Physical Protection Plan, compliant with applicable portions of 10 CFR Part 73, no later than 180 days after issuance of a construction authorization. The applicant's commitment is listed in SER Volume 1, Appendix.

In addition to describing how the protection goal will be met, the applicant stated that it will maintain and update the Physical Protection Plan to reflect changes necessary to ensure its continued effectiveness in meeting the protection goal. The NRC staff finds the applicant's description of its planned maintenance and updating of the Physical Protection Plan acceptable because this attribute is the same as that identified in the protection goal acceptance criterion.

Findings on Protection Goal

On the basis of the NRC staff's review of the applicant's information, and consistent with YMRP Section 1.3.3, Acceptance Criterion 3, the NRC staff makes the following findings.

On the basis of information provided in GI Section 3.3, and the applicant's commitment to submit to NRC a Physical Protection Plan, compliant with applicable portions of 10 CFR Part 73, no later than 180 days after the issuance of a construction authorization, the NRC staff finds that the protection goal will be met. The NRC staff finds that, with respect to a construction authorization, the applicant's description of its Physical Protection Plan acceptably addresses each of the attributes of the protection goal acceptance criterion.

Evaluation Findings

The NRC staff has reviewed the GI and other information submitted in support of the license application and finds, with reasonable assurance, that the requirements of 10 CFR 63.21(b)(3) are satisfied with respect to a construction authorization. DOE has described the detailed security measures for physical protection of HLW in accordance with 10 CFR 73.51(b)(3). DOE will design and submit plans for the physical protection system to protect against loss of control of the facility that could be sufficient to cause a radiation exposure exceeding the dose as described in 10 CFR 72.106.

1.3.3.2.4   Security Organization

In GI Section 3.4 the applicant described how the Physical Protection Plan addresses the security organization. The applicant provided information on the (i) employment status of the security organization (i.e., employed directly by DOE or is a contractor); (ii) staffing level of the security organization and its structure and management; (iii) review of the physical protection program; and (iv) training, equipping, qualifying, and requalifying members of the security organization. In GI Sections 3.6 and 3.12 the applicant provided information on the retention of records required by 10 CFR 73.51(d)(13).

The NRC staff finds the applicant description in GI Section 3.4 of the employment status of the security organization acceptable for two reasons. First, the applicant's description is acceptable because the applicant will indicate in the plan whether the security organization is employed directly by DOE or is a contractor to DOE. Second, the applicant's description is acceptable because the applicant will establish written agreements between DOE and the contract guard force that will govern how the contract security force meets the requirements of 10 CFR 73.51(d)(5) and 10 CFR Part 73, Appendix B. The NRC staff finds that these attributes described by the applicant are the same as those identified in the security organization acceptance criterion (YMRP Section 1.3.3, Acceptance Criterion 4, Subcriterion 1). Thus, the NRC staff finds the applicant's information on the employment status of the security organization, described in the Physical Protection Plan, acceptable.

The NRC staff finds the applicant description in GI Section 3.4 of the staffing level of the security organization, and the structure and management of the security organization, acceptable for two reasons. First, the applicant's description is acceptable because the applicant identified that the security organization will provide sufficient personnel for each shift to provide for monitoring of detection systems and the conduct of surveillance, assessments, access control, and communications to ensure adequate response. Second, the applicant's description is acceptable because the applicant identified that the plan will define security responsibilities and detail reporting lines from a Site Protection Manager (the Site Protection Manager's position is described in SAR Section 5.3.1.2.7) down to the employees who are assigned the responsibility for the direct supervision of physical security activities and security personnel. The NRC staff finds that these attributes described by the applicant in GI Section 3.4 are the same as those identified in the security organization acceptance criterion (YMRP Section 1.3.3, Acceptance Criterion 4, Subcriterion 2). Thus, the NRC staff finds the applicant's information on the staffing level of the security organization and its structure and management, described in the Physical Protection Plan, acceptable.

The NRC staff finds the applicant description in GI Section 3.4 of the review of the physical protection program acceptable for two reasons. First, the applicant's description is acceptable because the applicant identified how the physical protection program will be reviewed. The NRC staff finds the description acceptable because the applicant stated that, consistent with 10 CFR 73.51(d)(12), assessment of the Physical Protection Plan will be performed at least once every 24 months by individuals independent of both physical protection management and personnel who have direct responsibility for implementation. Second, the applicant's description is acceptable because the applicant identified what will be reviewed in its assessments. The NRC staff finds the description acceptable because the applicant stated that assessment will verify the effectiveness of the Physical Protection Plan and the liaison with and training of the designated offsite response force and any local law enforcement agency. The NRC staff finds that these applicant-described attributes, of what and how the physical protection program will be reviewed, are the same as those identified in the security organization acceptance criterion (YMRP Section 1.3.3, Acceptance Criterion 4, Subcriterion 3). Thus, the NRC staff finds the applicant's information on the review of the physical protection program, described in the Physical Protection Plan, acceptable.

In GI Section 3.4 the applicant described information that will be in the Physical Protection Plan on training and equipping the security organization and identified that the security force will operate under DOE authority to carry firearms and make limited arrests consistent with 10 CFR Part 1046 and 10 CFR Part 1047. The NRC staff finds the applicant description of the training, equipping, qualifying, and requalifying members of the security organization acceptable for two reasons. First, the applicant's description is acceptable because the applicant indicated that it would establish a security force training plan. Second, the applicant's description is acceptable because the applicant identified that the Physical Protection Plan will describe the process for selecting, qualifying, training, and equipping members of the security organization so they can perform their security duties as identified in the Physical Protection Plan, the Safeguards Contingency Plan, and the Training and Qualification Plan, in accordance with 10 CFR 73.51(d)(5). In addition, the applicant described other attributes (e.g., security force suitability and qualifications and firearms training) that will be addressed in the Physical Protection Plan and identified that the applicant will train, equip, qualify, and re-qualify members in accordance with 10 CFR Part 73, Appendix B. The NRC staff finds that these attributes described by the applicant are the same as those identified in the security organization acceptance criterion (YMRP Section 1.3.3, Acceptance Criterion 4, Subcriterion 4). Thus, the NRC staff finds the applicant's information on the training, equipping, qualifying, and requalifying members of the security organization, described in the Physical Protection Plan, acceptable.

In GI Sections 3.6 and 3.12 the applicant described its Physical Protection Plan approach for maintaining records as required by 10 CFR 73.51(d)(13). The NRC staff finds the applicant description of maintenance of records acceptable for the following reason. The applicant's description is acceptable because the applicant identified what records would be maintained and how long the records will be maintained. The NRC staff finds the description acceptable because the applicant identified that it will retain records for 3 years for the following items: (i) a log of individuals granted access to the protected area; (ii) screening records of members of the security organization; (iii) a log of all patrols; (iv) a record of each alarm received, identifying the type of alarm, location, date and time when received, and the disposition of the alarm; and (v) the Physical Protection Plan review reports. The NRC staff finds that the applicant's identification of what records will be maintained and for how long the records will be maintained are the same as those identified in the security organization acceptance criterion (YMRP Section 1.3.3, Acceptance Criterion 4, Subcriterion 5). Thus, the NRC staff finds the applicant's information on the review of the physical protection program, described in the Physical Protection Plan, acceptable.

Findings on Security Organization

On the basis of the NRC staff's review of the applicant's information, and consistent with YMRP Section 1.3.3, Acceptance Criterion 4, the NRC staff makes the following findings.

On the basis of information provided in GI Sections 3.4, 3.6, and 3.12, the NRC staff finds that the security organization will be adequate to manage, control, and implement the plan and to maintain its effectiveness. The NRC staff finds that, with respect to a construction authorization, the applicant's description of its Physical Protection Plan acceptably addressed each of the attributes of the security organization acceptance criterion.

Evaluation Findings

The NRC staff has reviewed the GI and other information submitted in support of the license application and finds, with reasonable assurance, that the requirements of 10 CFR 63.21(b)(3) are satisfied with respect to a construction authorization. DOE has described the detailed security measures for physical protection of HLW in accordance with 10 CFR 73.51(d)(5), (d)(12), and (d)(13); 10 CFR Part 73, Appendix B, Sections I.A.(1)(a) and (b) and B(1)(a); and the applicable portions of 10 CFR Part 73, Appendix B, Section II. DOE will establish a security organization with written procedures. DOE's security organization will include sufficient personnel per shift to provide for monitoring of detection systems and the conduct of surveillance, assessment access control, and communications to assure adequate response. Members of the DOE's security organization will be trained, equipped, qualified, and requalified to perform assigned job duties in accordance with 10 CFR Part 73, Appendix B, Sections I.A.(1)(a) and (b) and B(1)(a) and the applicable portions of 10 CFR Part 73, Appendix B, Section II. DOE's physical protection program will be reviewed once every 24 months by individuals independent of both physical protection program management and personnel who have direct responsibility for implementation of the physical protection program. The physical protection program review will include an evaluation of the effectiveness of the physical protection system and a verification of the liaison established with the designated response force and any local law enforcement agency. DOE will retain the following documentation as a record for 3 years after the record is made or until termination of the license: (i) a log of individuals granted access to the protected area; (ii) screening records of members of the security organization; (iii) a log of all patrols; (iv) a record of each alarm received, identifying the type of alarm, location, date and time when received, and disposition of the alarm; and (v) the physical protection program review reports.

1.3.3.2.5 Physical Barrier Subsystems

In GI Section 3.5 the applicant described how the Physical Protection Plan addresses physical barrier subsystems. The applicant provided information on the (i) physical barriers, (ii) isolation zones, and (iii) lighting system.

The NRC staff finds the applicant's description of the physical barriers in GI Section 3.5 acceptable for the following reasons. First, the applicant's description is acceptable because the applicant stated that the protected area of the GROA will be surrounded by physical barriers as defined in 10 CFR 73.2 and indicated that waste will be handled only within the protected area. Second, the applicant's description is acceptable because the applicant identified that access to nuclear material will require passage through or penetration of two physical barriers, one barrier at the perimeter of the protected area and one barrier offering substantial penetration resistance consistent with 10 CFR 73.51(d)(1). Third, the applicant's description is acceptable because the applicant identified that a physical barrier at the perimeter will be installed so that it cannot be lifted to allow an individual to crawl under it. Fourth, the applicant's description is acceptable because the applicant indicated that the barrier offering substantial resistance to penetration will be described in the Physical Protection Plan. Fifth, the applicant's description is acceptable because the applicant stated that the plan will describe access points through the protected area barrier, the manner in which access points are used, and the means to control and protect access to ensure the integrity of the barrier. The NRC staff finds that each of these five attributes the applicant described is the same as those identified in the physical barrier subsystem acceptance criterion (YMRP Section 1.3.3, Acceptance Criterion 5, Subcriterion 1). Finally, the applicant's description of the physical barrier subsystems in the Physical Protection Plan will address how the physical barrier systems will be modified as the various waste handling facilities are phased into operation. The NRC staff finds the applicant's description acceptable because the applicant's phased construction, described in GI Section 1.2.2, would require that the protected area within the GROA would change with time and the applicant will address this outcome from the phased construction approach within its plan. Thus, the NRC staff finds the applicant's information on the physical barrier subsystem, described in the Physical Protection Plan, acceptable.

The NRC staff finds the applicant's description of the isolation zones in GI Section 3.5 acceptable for three reasons. First, the applicant's description is acceptable because the applicant stated the physical barrier at the protected area perimeter will have at least 6.1-m [20-ft]-wide isolation zones on both sides of the barrier. Second, the applicant's description is acceptable because the applicant stated that the isolation zones will be clear of obstacles and structures to permit assessment consistent with 10 CFR 73.51(d)(1). Third, the applicant's description is acceptable because the Physical Protection Plan will describe the size and location of isolation zones. The NRC staff finds that these attributes the applicant described are the same as those identified in the physical barrier subsystem acceptance criterion (YMRP Section 1.3.3, Acceptance Criterion 5, Subcriterion 2). Thus, the NRC staff finds the applicant's information on the isolation zones, described in the Physical Protection Plan, acceptable.

The NRC staff finds the applicant's description of the lighting system in GI Section 3.5 acceptable for four reasons. First, the applicant's description is acceptable because the applicant stated the lighting system will permit assessment of unauthorized penetrations of or activities within the protected area, consistent with 10 CFR 73.51(d)(2). Second, the applicant's description is acceptable because the applicant identified that the lighting system will provide sufficient illumination for monitoring, observing, and assessing activities in exterior areas within the protected area. Third, the applicant's description is acceptable because the applicant stated that emergency backup power will be provided for security-specific vital equipment and select protected area lighting in case normal power is lost. Fourth, the applicant's description is acceptable because the applicant stated that illumination will be maintained during periods of darkness. The NRC staff finds that these four attributes the applicant described are the same as those identified in the physical barrier subsystem acceptance criterion (YMRP Section 1.3.3, Acceptance Criterion 5, Subcriterion 3). Thus, the NRC staff finds the applicant's information on the lighting systems, described in the Physical Protection Plan, acceptable.

Findings on Physical Barrier Subsystems

On the basis of the NRC staff's review of the applicant's information, and consistent with YMRP Section 1.3.3, Acceptance Criterion 5, the NRC staff makes the following findings.

On the basis of information provided in GI Section 3.5, the NRC staff finds that the physical barrier subsystems will be adequate. The NRC staff finds that, with respect to a construction authorization, the applicant's description of its Physical Protection Plan acceptably addressed each of the attributes of the physical barrier subsystems acceptance criterion.

Evaluation Findings

The NRC staff has reviewed the GI and other information submitted in support of the license application and finds, with reasonable assurance, that the requirements of 10 CFR 63.21(b)(3) are satisfied with respect to a construction authorization. DOE has described the detailed security measures for physical protection of HLW in accordance with 10 CFR 73.51(d)(1) and (d)(2). DOE will store SNF and HLW only within a protected area so that access to this material requires passage through or penetration of two physical barriers: one barrier at the perimeter of the protected area and one barrier offering substantial penetration resistance. DOE's physical barrier at the perimeter of the protected area will be defined as in 10 CFR 73.2. DOE's illumination will be sufficient to permit adequate assessment of unauthorized penetrations of or activities within the protected area.

1.3.3.2.6 Access Control Subsystems and Procedures

In GI Section 3.6 the applicant described how the Physical Protection Plan will address access control subsystems and procedures. The applicant provided information on the (i) goal for the access control subsystems and procedures, (ii) personnel identification system, (iii) procedures, (iv) lock control system, and (v) retention of access control records.

The NRC staff finds the applicant's description of the access control subsystems and procedures in GI Section 3.6 acceptable for two reasons. First, the applicant's description is acceptable because the applicant stated that controls and procedures will be developed and implemented to verify the identity of people, vehicles, and materials and to initiate timely response measures to deny unauthorized entries or material removal. The applicant's goal is consistent with the access control subsystems and procedures acceptance criterion (YMRP Section 1.3.3, Acceptance Criterion 6), and thus, the NRC staff finds it acceptable.

Second, the NRC staff also finds the applicant's description of access control subsystems and procedures acceptable because the applicant described the four access control subsystems that it will implement in its Physical Protection Plan, and these attributes are consistent with the access control subsystems and procedures acceptance criterion (YMRP Section 1.3.3, Acceptance Criterion 6, Subcriteria 1–4). The applicant first described that a personnel identification system, to be identified in the Physical Protection Plan, will be established and maintained to limit access to the protected area and the controlled access areas within it. The applicant also stated that the personnel identification system will provide for unique identification of individuals granted access to the controlled access area. The NRC staff finds that applicant's description of the personnel identification system is consistent with that identified in the YMRP Section 1.3.3, Acceptance Criterion 6, Subcriterion 1 because it described the same attributes. Next, the applicant described the procedures for controlling access. The applicant stated that procedures will (i) include appropriate methods of identifying individuals and verifying individual authorization, consistent with 10 CFR 73.51(d)(7); (ii) include techniques for conducting searches before entry into the protected area of individuals, vehicles, and hand-carried packages for explosives or other prohibited items that could be used for radiological sabotage, consistent with 10 CFR 73.51(d)(9); and (iii) provide protection against unauthorized removal of material, including theft. The NRC staff finds that these three attributes of the procedures for controlling access the applicant described are consistent with the acceptance criterion (YMRP Section 1.3.3, Acceptance Criterion 6, Subcriterion 2). Thus, the NRC staff finds the applicant's described procedures for controlling access acceptable. Next, the applicant's description of a lock control system identified that the lock system, consistent with 10 CFR 73.51(d)(7) and with the applicable guidance in Regulatory Guide 5.12 (NRC, 1973ac), will be a part of the physical barrier system. The NRC staff finds that the applicant's description of the lock system is consistent with the acceptance criterion (YMRP Section 1.3.3, Acceptance Criterion 6, Subcriterion 3), and thus, the NRC staff finds the applicant's description acceptable. Finally, the applicant described that records of access control will be retained, consistent with 10 CFR 73.51(d)(13), for 3 years after the record is generated or until NRC terminates the license. On the basis of the NRC staff's finding in SER Section 1.3.3.2.4 that the applicant's description of record retention is acceptable, the NRC staff finds the applicant's description of retention of access control records, relative to YMRP Section 1.3.3, Acceptance Criterion 6, Subcriterion 4, acceptable. Because the applicant described the four access control subsystems, consistent with the attributes of the acceptance criterion (YMRP Section 1.3.3, Acceptance Criterion 6), the NRC staff concludes that the applicant will provide adequate access control subsystems for the GROA.

Findings on Access Control Subsystems and Procedures

On the basis of the NRC staff's review of the applicant's information, and consistent with YMRP Section 1.3.3, Acceptance Criterion 6, the NRC staff makes the following findings.

On the basis of information provided in GI Section 3.6 and the NRC staff's finding in SER Section 1.3.3.2.4 on record retention, the NRC staff finds that the access control subsystems and procedures will be adequate. The NRC staff finds that, with respect to a construction authorization, the applicant's description of its Physical Protection Plan acceptably addressed each of the attributes of the access control subsystems and procedures acceptance criterion.

Evaluation Findings

The NRC staff has reviewed the GI and other information submitted in support of the license application and finds, with reasonable assurance, that the requirements of 10 CFR 63.21(b)(3) are satisfied with respect to a construction authorization. DOE has described the detailed security measures for physical protection of HLW in accordance with 10 CFR 73.51(d)(7), (d)(9), and (d)(13). DOE will establish and maintain a personnel identification system and a controlled lock system to limit access to authorized individuals. DOE will check all individuals, vehicles, and hand-carried packages entering the protected area for proper authorization and visually search for explosives before entry. DOE will retain a log of individuals granted access to the protected area as a record for 3 years after the record is made or until termination of the license.

1.3.3.2.7 Detection, Surveillance, and Alarm Subsystems and Procedures

In GI Section 3.7 the applicant described how the Physical Protection Plan will address detection, surveillance, and alarm subsystems and procedures. The applicant provided information on (i) the goal for these subsystems and procedures, (ii) characteristics of its active intrusion detection system, (iii) characteristics of its alarm system, (iv) tamper indication and operability of the intrusion detections systems and supporting subsystems, and (v) patrols of the protected area.

The NRC staff finds the applicant description in GI Section 3.7 of the detection, surveillance, and alarm subsystems and procedures acceptable for two reasons. First, the applicant's description is acceptable because the applicant stated that detection, surveillance, and alarm subsystems and implementing procedures will provide for real-time capabilities to detect, assess, and communicate any attempted unauthorized access or penetration by individuals, vehicles, or materials so the security force can prevent such access or penetration. The applicant's goal is consistent with the detection, surveillance, and alarm subsystems and procedures acceptance criterion, and thus, the NRC staff finds it acceptable.

Second, the NRC staff finds the applicant's description of detection, surveillance, and alarm subsystems and procedures acceptable because the applicant described the four attributes of its detection, surveillance, and alarm subsystems and procedures that it will implement in its Physical Protection Plan, and these attributes are consistent with the detection, surveillance, and alarm subsystems and procedures acceptance criterion. The applicant's description of these four attributes is individually evaluated in the following paragraphs.

The applicant stated that its active intrusion detection system will be at the protected area perimeter and will comply with the requirements of 10 CFR 73.51(d)(3) and with applicable guidance in Regulatory Guide 5.44 (NRC, 1997ad). The applicant's description also included these characteristics: (i) there will be no gaps in the coverage of the perimeter of the protected area; (ii) the perimeter of the protected area will be under continuous surveillance; and (iii) the protected area perimeter will be divided into multiple segments that are independently alarmed and monitored to assist the security force in assessing and responding to an alarm by localizing the area in which the alarm is initiated. The NRC staff finds that applicant's description of the active intrusion detection system is consistent with that identified in the acceptance criterion (YMRP Section 1.3.3, Acceptance Criterion 7, Subcriterion 1) because it described similar attributes and will comply with applicable guidance in Regulatory Guide 5.44 (NRC, 1997ad). Thus, the NRC staff finds the applicant's description of its active intrusion detection system acceptable.

The applicant stated that the Physical Protection Plan will describe the location, construction, and characteristics of the primary and secondary alarm stations consistent with 10 CFR 73.51(d)(3) and with applicable guidance in Regulatory Guide 5.44 (NRC, 1997ad). In GI Section 3.7 the applicant described some characteristics of its alarm system. The NRC staff finds that the characteristics the applicant described are consistent with the acceptance criterion and included (i) alarms that will annunciate in a continuously manned primary alarm station located within the protected area and in at least one additional, continuously staffed, independent, secondary alarm station to ensure that a single act cannot remove the capability of calling for assistance or responding to an alarm; (ii) access to the alarm stations will be controlled, and the primary alarm station functions will not include operational activities that could interfere with the execution of alarm response functions; and (iii) alarms indicating penetration, including the unauthorized opening of access points, will annunciate in both the primary and secondary alarm stations. Because the applicant described attributes that are the same as those identified in the acceptance criterion (YMRP Section 1.3.3, Acceptance Criterion 7, Subcriterion 2) and will comply with applicable guidance in Regulatory Guide 5.44 (NRC, 1997ad), the NRC staff finds the applicant's description of its alarm system acceptable.

The applicant described the tamper indication and operability of the intrusion detections systems and supporting subsystems. The applicant stated that, consistent with 10 CFR 73.51(d)(11) and applicable guidance in Regulatory Guide 5.44 (NRC, 1997ad), intrusion detection systems and supporting subsystems will be equipped with tamper-indicating devices and line supervision. The applicant identified that physical protection systems will be maintained in operable condition and timely compensatory measures will be implemented in the event of system outages. The NRC staff finds the applicant description acceptable because the described attributes are the same as those identified in the acceptance criterion (YMRP Section 1.3.3, Acceptance Criterion 7, Subcriterion 3) and will comply with applicable guidance in Regulatory Guide 5.44 (NRC, 1997ad). Combined with the NRC staff's finding in SER Section 1.3.3.2.9 that the applicant's description of equipment operability and compensatory measures is acceptable, the NRC staff finds the applicant's description of tamper indication and operability acceptable.

The applicant described its patrols of the protected area. The applicant identified that, consistent with 10 CFR 73.51(d)(4), the protected area will be monitored by daily, random patrols and that the number of patrols per shift will be addressed in the Physical Protection Plan or implementing procedures. The NRC staff finds the applicant's description of patrols of the protected area acceptable because the described attribute is the same as that in the acceptance criterion. Because the applicant described attributes that are the same as those identified in the acceptance criterion (YMRP Section 1.3.3, Acceptance Criterion 7, Subcriterion 4), the NRC staff finds the applicant's information on its patrols of the protected area acceptable.

<u>Findings on Detection, Surveillance, and Alarm Subsystems and Procedures</u>

On the basis of the NRC staff's review of the applicant's information, and consistent with YMRP Section 1.3.3, Acceptance Criterion 7, the NRC staff makes the following findings.

On the basis of information provided in GI Section 3.6 and the NRC staff's finding in SER Section 1.3.3.2.9 on equipment operability and compensatory measures, the NRC staff finds that the detection, surveillance, and alarm subsystems and procedures will be adequate. The NRC staff finds that, with respect to a construction authorization, the applicant's description of its Physical Protection Plan acceptably addressed each of the attributes of the detection, surveillance, and alarm subsystems and procedures acceptance criterion.

Evaluation Findings

The NRC staff has reviewed the GI and other information submitted in support of the license application and finds, with reasonable assurance, that the requirements of 10 CFR 63.21(b)(3) are satisfied with respect to a construction authorization. DOE has described the detailed security measures for physical protection of HLW in accordance with 10 CFR 73.51(d)(3), (d)(4), and (d)(11). DOE's perimeter of the protected area will be subject to continual surveillance and be protected by an active intrusion alarm system that is capable of detecting penetrations through the isolation zone and that is monitored in a continually staffed primary alarm station and in one additional continually staffed location. DOE's primary alarm station will be located within the protected area and have bullet-resisting walls, doors, ceiling and floor; the interior of the station will not be visible from outside the protected area. DOE will provide a timely means for assessment of alarms. DOE will monitor the protected area by daily random patrols. All of DOE's detection systems and supporting subsystems will be tamper indicating with line supervision. DOE's systems, as well as surveillance/assessment and illumination systems, will be maintained in operable condition. DOE will take timely compensatory measures after discovery of inoperability to assure that the effectiveness of the security system is not reduced.

1.3.3.2.8 Communication Subsystems

In GI Section 3.8 the applicant described how the Physical Protection Plan will address communication subsystems. The applicant stated in GI Section 3 that the communication subsystem is part of the physical protection system and described in GI Section 3.9 how equipment operability and compensatory measures for the physical protections systems will be addressed in the Physical Protection Plan.

The NRC staff finds the applicant's description of the communication subsystems in GI Section 3.8 acceptable for four reasons. First, the applicant's description is acceptable because the applicant identified that the communication subsystem will provide notification of attempted unauthorized intrusion into the protected area to the security force in each continuously manned alarm station. Second, the applicant's description is acceptable because the applicant identified that redundant systems will be provided to ensure the capability of communications between the security force and the designated offsite response force, in accordance with 10 CFR 73.51(d)(8). Third, the applicant's description is acceptable because the applicant identified attributes of the communication subsystem (e.g., personnel will be capable of calling for assistance and response forces, and primary and secondary alarm stations will be equipped to communicate with designated offsite response forces, including local law enforcement agencies). Fourth, the applicant's description is acceptable because the applicant identified that the communication subsystem will be maintained in operable condition. In addition, the applicant described in GI Section 3 that the communication subsystem is part of the physical protection system. In SER Section 1.3.3.2.9, the NRC staff finds the applicant's description of the equipment operability and compensatory measures for physical protection systems acceptably addressed the acceptance criterion (YMRP Section 1.3.3, Acceptance Criterion 9) and acceptably addressed 10 CFR 73.51(d)(11). Thus, the NRC staff finds that the applicant's methods used to maintain communications in operable condition will be adequate.

Findings on Communication Subsystems

On the basis of the NRC staff's review of the applicant's information, and consistent with YMRP Section 1.3.3, Acceptance Criterion 8, the NRC staff makes the following findings.

On the basis of information provided in GI Sections 3, 3.8, and 3.9 and the NRC staff's finding in SER Section 1.3.3.2.9 on equipment operability and compensatory measures, the NRC staff finds that the communication subsystems will be adequate. The NRC staff finds that, with respect to a construction authorization, the applicant's description of its Physical

Protection Plan acceptably addressed each of the attributes of the communication subsystem acceptance criterion.

Evaluation Findings

The NRC staff has reviewed the GI and other information submitted in support of the license application and finds, with reasonable assurance, that the requirements of 10 CFR 63.21(b)(3) are satisfied with respect to a construction authorization. DOE has described the detailed security measures for physical protection of HLW in accordance with 10 CFR 73.51(d)(8) and d(11). DOE will provide redundant communications capability between onsite security force members and designated response force or local law enforcement agencies. The communication subsystem will be maintained in operable condition.

1.3.3.2.9 Equipment Operability and Compensatory Measures

In GI Section 3.9 the applicant described how the Physical Protection Plan will address equipment operability and compensatory measures.

The NRC staff finds the applicant's description of the equipment operability and compensatory measures in GI Section 3.9 acceptable for two reasons. First, the applicant's description is acceptable because the applicant stated that tests and preventive maintenance procedures will provide confidence that security equipment is effective, available, reliable, and able to perform when needed, consistent with the requirements of 10 CFR 73.51(d)(11). Second, the applicant's description is acceptable because the applicant then identified the actions (e.g., implementing a test and maintenance program for physical protections systems) it will take to meet the stated purpose of providing confidence that security equipment will be available and reliable to perform when needed. For instance, the applicant stated that members of the security force will conduct patrols of the protected area daily at random intervals to verify the integrity of physical barriers, including movable openings such as gates. The applicant also stated that the testing program for the perimeter intrusion detection system will be consistent with applicable guidance in Regulatory Guide 5.44 (NRC, 1997ad). The applicant stated that it would verify the functional performance of lighting, security alarms, annunciators, and transmission to the primary alarm station. In addition the applicant described that it will take timely compensatory measures when integrity or performance measures are found to be deficient, in accordance with 10 CFR 73.51(d)(11). The NRC staff finds that these attributes the applicant described are consistent with those identified in the equipment operability and compensatory measures acceptance criterion (YMRP Section 1.3.3, Acceptance Criterion 9) and acceptably address 10 CFR 73.51(d)(11).

Findings on Equipment Operability and Compensatory Measures

On the basis of the NRC staff's review of the applicant's information, and consistent with YMRP Section 1.3.3, Acceptance Criterion 9, the NRC staff makes the following findings.

On the basis of information provided in GI Section 3.9, the NRC staff finds that the equipment operability and compensatory measures will be adequate. The NRC staff finds that, with respect to a construction authorization, the applicant's description of its Physical Protection Plan acceptably addressed each of the attributes of the equipment operability and compensatory measures acceptance criterion.

Evaluation Findings

The NRC staff has reviewed the GI and other information submitted in support of the license application and finds, with reasonable assurance, that the requirements of 10 CFR 63.21(b)(3) are satisfied with respect to a construction authorization. DOE has described the detailed security measures for physical protection of HLW in accordance with 10 CFR 73.51(d)(11). All of DOE's detection systems and supporting subsystems, as well as surveillance/assessment and illumination systems, will be maintained in operable condition. DOE will take timely compensatory measures after discovery of inoperability to assure that the effectiveness of the security system is not reduced.

1.3.3.2.10   Contingency and Response Plans and Procedures

In GI Section 3.10 the applicant described how the Physical Protection Plan will address contingency and response plans and procedures. The applicant provided information on (i) the goal for the contingency and response plans and procedures, (ii) its safeguards contingency plan, and (iii) documented response arrangements with a designated offsite response force.

The NRC staff finds the applicant description of the contingency and response plans and procedures in GI Section 3.10 acceptable for two reasons. First, the applicant's description is acceptable because the applicant stated that the Safeguards Contingency Plan and the implementing procedures will describe measures to provide predetermined responses to safeguards contingency events, so that an intruder will be engaged and impeded until offsite assistance arrives. The applicant's goal is consistent with the contingency and response plans and procedures acceptance criterion (YMRP Section 1.3.3, Acceptance Criterion 10), and thus, the NRC staff finds it acceptable.

Second, the NRC staff finds the applicant's description of contingency and response plans and procedures acceptable because the applicant described two attributes of its contingency and response plans and procedures that it will develop and maintain as part of its Physical Protection Plan, and these attributes are consistent with the contingency and response plans and procedures acceptance criterion (YMRP Section 1.3.3, Acceptance Criterion 10, Subcriteria 1–2). The applicant's description of the two attributes is individually evaluated in the following paragraphs.

First, the applicant described its Safeguards Contingency Plan. The applicant stated that a Safeguards Contingency Plan will be developed, maintained, and periodically reviewed and revised, consistent with the requirements of 10 CFR Part 73, Appendix C and will include response procedures incorporating a responsibility matrix as required by 10 CFR 73.51(d)(10) and 10 CFR Part 73, Appendix C. The applicant's description included four separate attributes. First, the applicant described that as required by 10 CFR 73.51(d)(10), the Safeguards Contingency Plan will be maintained and updated, as necessary, until NRC terminates the license. Second, the applicant described that if any portion of the Safeguards Contingency Plan is superseded, the superseded portion will be kept on file for 3 years after the effective date of the change or until termination of the license. Third, the applicant described that the Safeguards Contingency Plan will identify specific objectives in the event of threats, theft, or radiological sabotage. Fourth, the applicant described that Safeguards Contingency Plan will specify the actions to be taken by the security force at the GROA and by repository management. The NRC staff finds the applicant description of the Safeguards Contingency Plan and its attributes acceptable because it is consistent with the contingency and response plans and procedures acceptance criterion (YMRP Section 1.3.3, Acceptance Criterion 10,

Subcriterion 1). Thus, the NRC staff finds that the applicant's information on the contingency and response plans and procedures, described in the Physical Protection Plan, is acceptable.

Second, the applicant described its documented response arrangements with a designated offsite response force. The applicant stated that the documented response arrangements will be made with a designated offsite response force, consistent with the requirements of 10 CFR 73.51(d)(6). The applicant also identified that if the designated offsite response force is privately contracted, it will meet the requirements of 10 CFR Part 73, Appendix B. Because the applicant's description of documented response arrangements with a designated offsite response force is consistent with the contingency and response plans and procedures acceptance criterion (YMRP Section 1.3.3, Acceptance Criterion 10, Subcriterion 2), the NRC staff finds that the applicant's description acceptable. Thus, the NRC staff finds the applicant's information on the contingency and response plans and procedures, described in the Physical Protection Plan, is acceptable.

Findings on Contingency and Response Plans and Procedures

On the basis of the NRC staff's review of the applicant's information, and consistent with YMRP Section 1.3.3, Acceptance Criterion 10, the NRC staff makes the following findings.

On the basis of information provided in GI Section 3.10, the NRC staff finds that the contingency and response plans and procedures will be adequate. The NRC staff finds that, with respect to a construction authorization, the applicant's description of its Physical Protection Plan acceptably addressed each of the attributes of the contingency and response plans and procedures acceptance criterion.

Evaluation Findings

The NRC staff has reviewed the GI and other information submitted in support of the license application and finds, with reasonable assurance, that the requirements of 10 CFR 63.21(b)(3) are satisfied with respect to a construction authorization. DOE has described the detailed security measures for physical protection of HLW in accordance with 10 CFR 73.51(d)(6) and (d)(10) and 10 CFR Part 73, Appendices B and C. DOE will establish and document liaison with a designated response force or local law enforcement agency to permit timely response to unauthorized penetration or activities. If the designated offsite response force is privately contracted, DOE will ensure the offsite response force meets the requirements of 10 CFR Part 73, Appendix B. DOE will establish and maintain written response procedures for addressing unauthorized penetration of or activities within, the protected area including those outlined in 10 CFR Part 73, Appendix C, Category 5, "Procedures." DOE will retain a copy of response procedures as a record for 3 years until termination of the license and will retain copies of superseded material for 3 years after each change or until termination of the license.

1.3.3.2.11   Reporting of Safeguards Events

In GI Section 3.11 the applicant described the Physical Protection Plan approach for reporting of safeguards events.

The NRC staff finds that the applicant's description stated that safeguards events will be reported to the NRC, consistent with the requirements of 10 CFR Part 73, Appendix G. The applicant described that the Physical Protection Plan will identify those events that are required to be reported to the NRC within 1 hour of discovery, followed by a written report within 60 days.

The applicant's description of the plan indicated that the plan will also require identification of events that will be recorded in the safeguards log within 24 hours of discovery. Because the applicant stated that safeguards events will be reported to NRC, consistent with the requirements of 10 CFR Part 73, Appendix G, and described disposition of the two classes of safeguard events consistent with Appendix G, the NRC staff finds the applicant's description of reporting of safeguards events acceptable. The attributes that the applicant described constitute adequate reporting of safeguards events and thus acceptably addressed the reporting of safeguards events acceptance criterion (YMRP Section 1.3.3, Acceptance Criterion 11).

Findings on Reporting of Safeguards Events

On the basis of the NRC staff's review of the applicant's information, and consistent with YMRP Section 1.3.3, Acceptance Criterion 11, the NRC staff makes the following findings.

On the basis of information provided in GI Section 3.11, the NRC staff finds that reporting of safeguards events will be adequate. The NRC staff finds that, with respect to a construction authorization, the applicant's description of its Physical Protection Plan acceptably addressed each of the attributes of the reporting of safeguards acceptance criterion.

Evaluation Findings

The NRC staff has reviewed the GI and other information submitted in support of the license application and finds, with reasonable assurance, that the requirements of 10 CFR 63.21(b)(3) are satisfied with respect to a construction authorization. DOE has described the detailed security measures for physical protection of HLW in accordance with 10 CFR Part 73, Appendix G. DOE will report safeguards events to NRC consistent with the requirements of 10 CFR Part 73, Appendix G.

1.3.3.2.12   Findings on Physical Protection Plan

On the basis of the information the applicant provided and the preceding review, and consistent with 10 CFR 63.21(b)(3), the NRC staff makes the following findings.

The NRC staff finds that the applicant's description of the detailed security measures for physical protection of SNF and HLW at the Yucca Mountain repository is complete and acceptably addresses each of the acceptance criteria in YMRP Section 1.3.3. Therefore, the NRC staff finds the applicant's description of its plan acceptable.

The applicant's commitment to submit to NRC a Physical Protection Plan, compliant with applicable portions of 10 CFR Part 73, no later than 180 days after the issuance of a construction authorization, is acceptable because the applicant submitted a description of its physical protection plan, which acceptably addressed each of the attributes of the reporting of safeguards acceptance criteria, and which included a commitment to submit a detailed Physical Protection Plan, compliant with applicable portions of 10 CFR Part 73, at a later date. 10 CFR 63.21(b)(3) requires DOE to only submit a description of the physical protection plan. Therefore DOE's submission is consistent with 10 CFR Part 63, because it includes a description of the detailed security measures for physical protection of high-level radioactive waste in accordance with 10 CFR 73.51, and generally described the design for physical protection, DOE's safeguards contingency plan, security organization personnel training and qualification plan, how the physical protection system is performance-tested to provide assurance that the system functions as intended, and how the system is tested and maintained to ensure its continued effectiveness, reliability, and availability. The applicant's commitment is listed in SER Volume 1, Appendix.

Based on its review in SER Section 1.3.3.2.1, the NRC staff finds that the applicant's security program will be operational and performance-tested prior to receipt of SNF and HLW.

1.3.4 Evaluation Findings

The NRC staff has reviewed the GI and other information submitted in support of the license application and finds, with reasonable assurance, that the requirements of 10 CFR 63.21(b)(3) are satisfied with respect to a construction authorization. DOE will implement a physical protection program for SNF and HLW that includes physical protection, a safeguards contingency plan, and a security organization personnel training and qualification plan that complies with 10 CFR 73.51.

1.3.5 References

DOE. 2008ab. DOE/RW–0573, "Yucca Mountain Repository License Application." Rev. 0. ML081560400. Las Vegas, Nevada: DOE, Office of Civilian Radioactive Waste Management.

NRC. 2007ae. "Geologic Repository Operations Area Security and Material Control and Accounting Requirements: Proposed Rule." *Federal Register.* Vol. 72, No. 244. pp. 72522–72562. Washington, DC: NRC.

NRC. 2003aa. NUREG–1804, "Yucca Mountain Review Plan—Final Report." Rev. 2. Washington, DC: NRC.

NRC. 1997ad. Regulatory Guide 5.44, "Perimeter Intrusion Alarm Systems." Rev. 3. Washington, DC: NRC.

NRC. 1973ac. Regulatory Guide 5.12, "General Use of Locks in the Protection and Control of Facilities and Special Nuclear Materials." Washington, DC: NRC.

CHAPTER 4

1.4 Material Control and Accounting Program

1.4.1 Introduction

This chapter of the Safety Evaluation Report (SER) evaluates the U.S. Department of Energy's (DOE or applicant) information on material control and accounting. The U.S. Nuclear Regulatory Commission (NRC) staff's evaluation is based on information provided in General Information (GI) Section 4 (DOE, 2008ab). GI Section 4 contains the applicant's description of the measures for material control and accounting of spent nuclear fuel (SNF) and high-level radioactive waste (HLW) at the Yucca Mountain repository. The applicant's information on material control and accounting of SNF and HLW at the repository at Yucca Mountain is part of the information that the NRC staff will consider, once the NRC staff's evaluation is completed for all volumes of the SER, to determine whether the common defense and security requirements, for construction authorization, have been met.

SER Volume 1, Chapters 1–3, provide information that allows the NRC staff to determine, in this chapter, whether the applicant's information on its material control and accounting program is consistent with the applicant's overview of its engineering design concept, schedule, and Physical Protection Program for the repository.

1.4.2 Regulatory Requirements

The requirements for the description of the measures for material control and accounting of SNF and HLW at the Yucca Mountain repository are in 10 CFR 63.21(b)(4), which states that the general information must include a description of the material control and accounting program to meet the requirements of 10 CFR 63.78. 10 CFR 63.78 requires that DOE shall implement a program of material control and accounting (and accidental criticality reporting) that is the same as that specified in 10 CFR 72.72, 72.74, 72.76, and 72.78.

10 CFR 72.72 specifies the material balance, inventory, and record requirements for stored materials. 10 CFR 72.72(a) requires that the applicant keep records showing the receipt, inventory (including location), disposal, acquisition, and transfer of all special nuclear material with quantities as specified in 10 CFR 74.13(a). 10 CFR 72.72 also specifies the requirements for (i) physical inventories in 10 CFR 72.72(b); (ii) the establishment, maintenance, and use of material control and accounting procedures in 10 CFR 72.72(c); and (iii) records in 10 CFR 72.72(d). 10 CFR 72.74 specifies the requirements for reporting accidental criticality or loss of special nuclear material. 10 CFR 72.76(a) identifies the requirements for material status reports, including (i) the requirement that each licensee shall complete in computer-readable format and submit to the Commission a Material Balance Report and a Physical Inventory Listing Report as specified in the instructions in NUREG/BR–0007 (NRC, 2003ac) and Nuclear Materials Management and Safeguards System (NMMSS) Report D–24 "Personal Computer Data Input for NRC Licensees" and (ii) that these reports, as specified by 10 CFR 74.13, provide information concerning the special nuclear material possessed, received, transferred, disposed of, or lost by the licensee. 10 CFR 72.78 specifies the requirements for nuclear material transaction reports, including the requirement that whenever the licensee transfers or receives or adjusts the inventory, in any manner, of special nuclear material as specified by 10 CFR 74.15, the licensee shall complete in computer-readable format a Nuclear Material Transaction Report as specified in the

instructions in NUREG/BR–0006 (NRC, 2003ad) and NMMSS Report D–24, "Personal Computer Data Input for NRC Licensees."

Because, at this time, the U.S. Government has not yet identified the Yucca Mountain facility as one of the eligible facilities under the "Agreement Between the United States of America and the International Atomic Energy Agency for the Application of Safeguards in the United States of America" (USA/IAEA, 1977aa), and because the applicant has not conducted any reportable activities identified in the "Protocol Additional to the Agreement Between the United States of America and the International Atomic Energy Agency for the Application of Safeguards in the United States of America" (USA/IAEA, 1998aa), the applicant is not yet subject to any of the requirements of 10 CFR 63.47 or 10 CFR 74.51.

On December 20, 2007, NRC published a proposed rule (NRC, 2007ae) to revise material control accounting requirements for the geologic repository operations area (GROA). The proposed rule would establish general performance objectives and corresponding system capabilities for the GROA material control and accounting program, with a focus on strengthening, streamlining, and consolidating all material control and accounting regulations specific to a GROA, and would affect portions of 10 CFR Parts 63 and 74 that apply to the Yucca Mountain repository. Because the proposed rule has not been finalized at the time of the NRC staff's review of the license application the applicant was not required to address the proposed changes. However, as described in SER Section 1.4.3.1, DOE has committed to submit a Material Control and Accounting Program, compliant with applicable portions of 10 CFR Part 74, no later than 180 days after the issuance of a construction authorization. The NRC staff will review the adequacy of DOE's Material Control and Accounting Program consistent with applicable requirements of 10 CFR Parts 63 and 73.

The NRC staff reviewed the applicant's material control and accounting program using the guidance in the Yucca Mountain Review Plan (YMRP) Section 1.4 (NRC, 2003aa). As described in YMRP Section 1.4, at the construction authorization stage DOE is required to submit a description of the material control and accounting program to meet the requirements of 10 CFR 63.78. YMRP Section 1.4.3 identifies four criteria that the NRC staff may consider in its evaluation:

1. Material balance, inventory, and record-keeping procedures for SNF and HLW are adequate.

2. Procedures are adequate to ensure timely reports of accidental criticality or loss of special nuclear material.

3. Procedures for preparation of material status reports are adequate.

4. Procedures for preparation of nuclear material transfer reports are adequate.

1.4.3 Technical Review

In SER Section 1.4.3.1 the NRC staff summarizes the applicant's license application for the Material Control and Accounting Program (the formal name of the applicant's program for material control and accounting). In SER Section 1.4.3.2 the NRC staff documents its review relative to the YMRP Section 1.4.3 acceptance criteria and summarizes its review methodology. The NRC staff's evaluation in subsequent separate sections (SER Sections 1.4.3.2.1–1.4.3.2.4) corresponds to the individual topics of the YMRP Section 1.4.3

acceptance criteria. In SER Section 1.4.3.2.5 the NRC staff summarizes its evaluation and presents its findings.

1.4.3.1 Summary of the DOE License Application on Material Control and Accounting Program

In GI Section 4, the applicant provided a description of the Material Control and Accounting Program, which the applicant stated meets the requirement of 10 CFR 63.78 and incorporates the requirements contained in applicable portions of 10 CFR Part 74. The applicant committed to submit to NRC a Material Control and Accounting Program, compliant with applicable portions of 10 CFR Part 74, no later than 180 days after issuance of a construction authorization. The applicant identified that the Material Control and Accounting Program will (i) include design basis information and assess potential impacts of the program on design features; (ii) use written procedures to account for and control SNF and HLW until NRC terminates the license and the repository is closed; and (iii) describe, establish, implement, and maintain procedures to protect against, detect, and respond to the potential loss of SNF and HLW, including loss through possible theft or diversion. In GI Section 4 the applicant summarized how the Material Control and Accounting Program will be designed and maintained to provide (i) accurate and current knowledge of SNF and HLW at the repository, (ii) annual confirmation of special nuclear material inventory, (ii) a collusion protection program, (iv) actions it will take for indications of missing special nuclear material, and (v) reporting of accidental criticality. The applicant presented information on (i) material balance, inventory, and record-keeping procedures in GI Section 4.1; (ii) reports of accidental criticality or loss of special nuclear material in GI Section 4.2; (iii) material status reports in GI Section 4.3; and (iv) nuclear material transfer reports in GI Section 4.4

1.4.3.2 NRC Staff Evaluation of Material Control and Accounting Program

In the following sections the NRC staff evaluates the applicant's description of its Material Control and Accounting Program. In conducting its review, the NRC staff compares the applicant's description to all the subcriteria of each acceptance criterion to ensure that the applicant's description of its Material Control and Accounting Program, in support of a construction authorization, is complete and consistent with the guidance in YMRP Section 1.4

1.4.3.2.1 Material Balance, Inventory, and Record-Keeping Procedures

YMRP Section 1.4.3 identifies seven subcriteria for Acceptance Criterion 1. The NRC staff evaluates the applicant's description relative to each of the subcriteria of Acceptance Criterion 1 in SER Sections 1.4.3.2.1.1–1.4.3.2.1.7. The NRC staff summarizes its evaluation of the applicant's information on material balance, inventory, and record-keeping procedures in SER Section 1.4.3.2.1.8.

1.4.3.2.1.1 Material Control and Accounting Plan

In GI Sections 4.1.1, 4.1.4, and 4.1.5 the applicant provided information on program provisions and requirements, periodic program assessment, and procedures for its Material Control and Accounting Program. The NRC staff verifies that the applicant described in GI Section 4.1.1, and consistent with 10 CFR 72.72(a), how the Material Control and Accounting Program will ensure that the material balance, inventory, and record-keeping procedures for HLW and SNF are implemented and effectively managed. The applicant described that the design for the Material Control and Accounting Program will be based upon the design of the repository, including the Physical Protection Plan discussed in GI Section 3, and that the material control and accounting features will be integrated, as appropriate, with repository design features and operations. On the basis of the NRC staff's evaluation of the applicant's information on (i) General Description; (ii) Proposed Schedules for Construction, Receipt, and Emplacement of Waste; and (iii) Physical Protection, reviewed in SER Sections 1.1, 1.2, and 1.3, respectively, the NRC staff finds the applicant's description of its Material Control and Accounting Program is consistent with the applicant's overview of its engineering design concept, schedule, and Physical Protection Program for the repository. The applicant has committed to submit to NRC both the Physical Protection Plan and its Material Control and Accounting Program no later than 180 days after issuance of a construction authorization. The applicant's commitments are listed in SER Volume 1, Appendix.

The applicant's commitment to submit to the NRC its Material Control and Accounting Program, compliant with applicable portions of 10 CFR Part 74, no later than 180 days after issuance of a construction authorization, is acceptable because, as described in SER Section 1.4.3.2.5, the applicant submitted a description of its Material Control and Accounting Program, which acceptably addressed each of the material control and accounting acceptance criteria, and a commitment to submit a detailed Material Control and Accounting Program, compliant with applicable portions of 10 CFR Part 74, at a later date. 10 CFR 63.21(b)(4) requires DOE to submit only a description of the material control and accounting program to meet the requirements of 10 CFR 63.78. The described material control and accounting program includes: (i) a description of material balance, inventory, and record-keeping procedures; (ii) reports of accidental criticality or loss of special nuclear material; (iii) material status reports; and (iv) nuclear material transaction reports.

The NRC staff verifies that the applicant described, in GI Section 4.1.1, that it will account for and control special nuclear material received at the repository, as a component of SNF and HLW, through receipt, processing, aging, and emplacement, until closure of the repository. Also, the NRC staff verifies that the applicant identified in the event that SNF or HLW is transferred from the repository, the applicant stated it will account for such transfer. The NRC staff verifies that the applicant described the actions that it will take, including (i) determine special nuclear material quantities associated with receipt of SNF and HLW transportation casks; (ii) maintain an item control program for identifying and tracking items (e.g., casks, individual fuel elements) containing special nuclear material; (iii) use tamper-indicating devices, as well as enhanced access control to strategic special nuclear material that meets the NRC definition of Category IA or IB material, as defined by 10 CFR 74.4; and (iv) establish a record-keeping system. The NRC staff verifies that the applicant described, in GI Section 4.1.4, that procedures will require that assessments of the Material Control and Accounting Program be performed no less frequently than 24 months between any 2 consecutive assessments to ensure the quality of material balances, physical inventories, and the effectiveness of the overall program. Also, the NRC staff verifies that the applicant described, in GI Section 4.1.5, that material control and accounting procedures will be established, tested, maintained, and followed.

The NRC staff finds that each of the applicant's described attributes of the material control and accounting plan, within its Material Control and Accounting Program, is consistent with YMRP Section 1.4.3, Acceptance Criterion 1, Subcriterion 1, and thus the applicant's license application, regarding its material control and accounting plan, acceptably addresses Acceptance Criterion 1.

1.4.3.2.1.2   Record Information

In GI Section 4.1.7 the applicant described records within its Material Control and Accounting Program. The NRC staff verifies that the applicant described that records for waste receipt, inventory (including location), disposal, acquisition, and transfer of SNF and HLW, including maintenance of inventory during any retrieval operations, will be documented. The NRC staff confirms that the applicant described that the records will provide information about the waste form, waste package, characteristics of any encapsulation material, radionuclide characteristics, heat generation rate, and history of the waste form. The NRC staff confirms that the applicant described that records will be maintained onsite from the time that the material arrives at the repository until 5 years after the repository is closed. The NRC staff verifies that the applicant described that information in the retained records will include (i) name of shipper; (ii) estimated quantity of radioactive material per item, including HLW; (iii) item identification and seal number; (iv) aging or emplacement location; (v) onsite movement of each fuel assembly or waste form canister; and (vi) ultimate disposal. The NRC staff finds the applicant's use of the term aging, instead of storage as used in YMRP Section 1.4.3, Acceptance Criterion 1, Subcriterion 2, acceptable when describing storage or emplacement location for the following reason. The applicant's usage of aging is appropriate because (i) in GI Section 1.1.2.1 the applicant identified the aging facilities are part of the surface GROA and are where aging of SNF occurs and (ii) in SER Sections 1.1.3.2.1 and 1.1.3.2.2 the NRC staff found the applicant's general description of the location and arrangement of the GROA, and general nature of the activities to be conducted at the GROA, respectively, acceptable. The NRC staff finds that each of the applicant's described attributes of records, within its Material Control and Accounting Program, is consistent with YMRP Section 1.4.3, Acceptance Criterion 1, Subcriterion 2, and thus the applicant's license application, regarding its records, acceptably addresses Acceptance Criterion 1.

1.4.3.2.1.3   Physical Inventory

In GI Section 4.1.2 the applicant described item accounting and physical inventories. The NRC staff confirms that the applicant described that, consistent with 10 CFR 72.72(b), DOE will conduct physical inventories at intervals not to exceed 12 months, unless directed otherwise by NRC, to account for all SNF and HLW containing special nuclear material at the repository. The NRC staff finds that the applicant's description of physical inventory, within its Material Control and Accounting Program, is consistent with YMRP Section 1.4.3, Acceptance Criterion 1, Subcriterion 3, and thus the applicant's license application, regarding its physical inventory, acceptably addresses Acceptance Criterion 1.

1.4.3.2.1.4   Quality of Physical Inventories

In GI Section 4.1.3 the applicant described quality of physical inventories. The NRC staff verifies that the applicant described that the Material Control and Accounting Program will require that policies, practices, and procedures be designed and implemented to ensure the quality of physical inventories and the control and maintenance of records and documentation associated with the physical inventories. The NRC staff verifies that the applicant described in GI Section 4.1.7 that consistent with 10 CFR 72.72(b), a copy of each current inventory will be retained until NRC terminates the license. The NRC staff finds that each of the applicant's described attributes of the quality of its physical inventories, within its Material Control and Accounting Program, is consistent with YMRP Section 1.4.3, Acceptance Criterion 1, Subcriterion 4, and thus the applicant's license application, regarding its quality of physical inventories, acceptably addresses Acceptance Criterion 1.

1.4.3.2.1.5   Procedures

In GI Section 4.1.5 the applicant described its material control and accouting procedures. The NRC staff confirms that the applicant described that it will establish, test, maintain, and follow material control and accounting procedures. Also, the NRC staff confirms that the applicant identified that, consistent with 10 CFR 72.72(c), copies of material control and accounting procedures used to conduct physical inventories will be maintained as part of the inventory record until NRC terminates the license. The NRC staff finds that the applicant's description of its procedures, within its Material Control and Accounting Program, is consistent with YMRP Section 1.4.3, Acceptance Criterion 1, Subcriterion 5, and thus the applicant's license application, regarding its procedures, acceptably addresses Acceptance Criterion 1.

1.4.3.2.1.6   Detection of Falsification of Data and Reports

In GI Section 4.1.6 the applicant described detection of falsification of data and reports. The NRC staff verifies that the applicant described that its Material Control and Accounting Program will have a collusion protection program to thwart attempts by an insider to divert SNF and HLW. The NRC staff also verifies that the applicant described that its collusion protection program will be designed to (i) provide confidence in the integrity of the traceability of the item accounting methods, (ii) detect falsification of data and reports, and (iii) protect against theft or diversion by insiders acting individually or in collusion. The NRC staff identifies that the applicant described that the collusion protection program will include procedures that provide for use of (i) a two-person rule when records are created for documenting item identification and tamper-indicating devices; (ii) redundant and separate transaction records; (iii) a two-person rule when nuclear material is moved to or from assigned locations; (iv) identification and elimination of possible diversion pathways through conduct of diversion-path analyses; (v) a personnel qualifications process; (vi) access control; and (vii) material control and accounting system performance testing. The NRC staff finds that the applicant's description of detection of falsification of data and reports, within its Material Control and Accounting Program, is consistent with YMRP Section 1.4.3, Acceptance Criterion 1, Subcriterion 6, and thus the applicant's license application, regarding detection of falsification of data and reports, acceptably addresses Acceptance Criterion 1.

1.4.3.2.1.7   Records Storage

The NRC staff confirms that the applicant described, in GI Section 4.1.7, that it will maintain duplicate sets of records at separate locations so that a single event does not destroy both sets of records. Also, the NRC staff confirms that the applicant described that records of any transfer of SNF or HLW out of the repository will be preserved for a minimum of 5 years after transfer in accordance with 10 CFR 72.72(d). The NRC staff finds that each of the applicant's described attributes of records storage, within its Material Control and Accounting Program, is consistent with YMRP Section 1.4.3, Acceptance Criterion 1, Subcriterion 7, and thus the applicant's license application, regarding records storage, acceptably addresses Acceptance Criterion 1.

1.4.3.2.1.8   Summary of NRC Staff's Evaluation of Material Balance, Inventory, and Record-Keeping Procedures

In SER Sections 1.4.3.2.1.1–1.4.3.2.1.7, the NRC staff evaluation finds that each of the applicant's described attributes of its material balance, inventory, and record-keeping procedures is consistent with, or the same as, the seven subcriteria presented in

YMRP Section 1.4.3, Acceptance Criterion 1. In SER Sections 1.4.3.2.1.1–1.4.3.2.1.7 the NRC staff concludes that the applicant's license application acceptably addresses Acceptance Criterion 1. Therefore, the NRC staff finds that, in support of a construction authorization, the applicant's description of material balance, inventory, and record-keeping procedures for SNF and HLW is adequate.

1.4.3.2.2   Reports of Accidental Criticality or Loss of Special Nuclear Material

In GI Sections 4.1.6 and 4.2 the applicant provided information on reports of accidental criticality or loss of special nuclear material within its Material Control and Accounting Program. As part of its description on reporting of loss of special nulcear material in GI Section 4.1.6, the applicant described its collusion protection program (a program that thwarts attempts by an insider to divert SNF and HLW). The NRC staff verifies the applicant described that its Material Control and Accounting Program will have a collusion protection program to thwart attempts by an insider to divert SNF and HLW. The NRC staff also verifies that the applicant described the design of its collusion protection program (e.g., to detect falsification of data and reports) and procedures of its collusion protection program (e.g., that provide for the use of a two-person rule when nuclear material is moved to or from assigned locations). The NRC staff finds that the applicant's described attributes of its collusion protection program within its Material Control and Accounting Program are consistent with YMRP Section 1.4.3, Acceptance Criterion 2, Subcriterion 1.

In GI Section 4.2 the applicant described its material control and accounting procedures that identify documentation requirements for reporting, investigating, and resolving missing special nuclear material and reporting of accidental criticality events within its Material Control and Accounting Program. The NRC staff verifies that the applicant described that (i) its reporting procedures will require that any potential anomalies be reported to NRC; (ii) potential anomalies include, but are not limited to, indicators at alarm levels such as those reflecting off-normal or situations; (iii) such anomalies could suggest a likelihood that special nuclear material may be missing, whether or not the cause is deliberate; and (iv) potential anomalies that would initiate procedures for reporting include, but are not limited to, missing items, falsified special nuclear material records, and violation of the two-person rule.

The NRC staff confirms that the applicant described that the anomaly reporting system will enable prompt response to alarms indicating a potential loss of special nuclear material and willl allow determination of whether the unusual observable condition is caused by an actual loss or by a system error. The NRC staff verifies that the applicant described that its reporting and resolution process will identify the anomaly and cause, so remedial action can be taken. Also, the NRC staff verifies that the applicant described that its response will be timely to ensure that indicators that might result from diversion, loss, or other misuse are investigated and resolved promptly.

The NRC staff confirms that the applicant described that it will develop and maintain, through its Material Control and Acounting Program, procedures for reporting accidental criticality or loss of special nuclear material to the NRC Operations Center, using the Emergency Notification System, within one 1 hour, in accordance with 10 CFR 72.74. Also, the NRC staff confirms that the applicant identified that if the Emergency Notification System is inoperable or unavailable, the required notification will be made by commercial telephone or any means that ensures that the NRC Operations Center receives a report within 1 hour of discovery.

The NRC staff finds that each of the applicant's described attributes of reports of accidental criticality or loss of special nuclear material within its Material Control and Accounting Program is consistent with, or the same as, the subcriteria presented in YMRP Section 1.4.3, Acceptance Criterion 2, and thus the applicant's license application acceptably addresses Acceptance Criterion 2. Therefore, the NRC staff finds that the applicant's described procedures are adequate to ensure timely reports of accidental criticality or loss of special nuclear material.

1.4.3.2.3 Material Status Reports

In GI Section 4.3 the applicant described material status reports within its Material Control and Accounting Program. The NRC staff verifies the applicant described that procedures will be developed to require that a material status report, in computer-readable format, will be completed by its material control and accounting staff and submitted to NRC, in accordance with the instructions in NUREG/BR–0007 (NRC, 2003ac) and Personal Computer Data Input for NRC Licensees (Nuclear Assurance Corporation International, 2001aa). The NRC staff confirms the applicant described that it will report information on the amount of special nuclear material it possessed, received, transferred, disposed of, or lost. The NRC staff finds that each of the applicant's described attributes of the material status reports within its Material Control and Accounting Program, evaluated in this paragraph, is consistent with the applicable subcriterion presented in YMRP Section 1.4.3, Acceptance Criterion 3.

The NRC staff confirms the applicant described that in accordance with 10 CFR 72.76(a), a material status report will be filed within 60 days of beginning the physical inventory required by 10 CFR 72.72(b), unless specified otherwise by the NRC. YMRP Section 1.4.3, Acceptance Criterion 3, states that procedures require material status reports as of March 31 and September 30 of each year, to be filed within 30 days after the end of the period covered by the report, unless otherwise specified by NRC or by 10 CFR 75.35, pertaining to implementation of the USA/IAEA Safeguards Agreement. However, because the Yucca Mountain repository is not subject to the requirements of 10 CFR 74.51 or 10 CFR 63.47 at this time, this statement in the YMRP is not applicable. If, in the future, the Yucca Mountain repository became subject to the requirements of 10 CFR 74.51, or 10 CFR 63.47, DOE would need to address the different reporting requirements. Thus, the NRC staff finds that each of the applicant's described attributes of the material status reports within its Material Control and Accounting Program is consistent with the applicable subcriterion presented in YMRP Section 1.4.3, Acceptance Criterion 3, and thus the applicant's license application acceptably addresses Acceptance Criterion 3. Therefore, the NRC staff finds that, in support of a construction authorization, the applicant's described procedures for preparation of material transfer reports are adequate.

1.4.3.2.4 Nuclear Material Transfer Reports

In GI Section 4.4 the applicant described nuclear material transfer reports within its Material Control and Accounting Program. The NRC staff confirms that the applicant described that auditable records pertaining to receipt and disposal of SNF and HLW, sufficient to demonstrate that reporting requirements have been met, will be developed and retained until NRC terminates the license. The NRC staff verifies the applicant identified that material control and accounting procedures will specify the form in which those records are kept. The NRC staff verifies the applicant described that its material control and accounting procedures will provide safeguards against tampering with or the loss of records. The NRC staff verifies that the applicant described that, consistent with 10 CFR 72.78(a), its material control and accounting procedures will require that, whenever special nuclear material is transferred or received, a nuclear material transaction report is completed in computer-readable format, in accordance with the instructions in NUREG/BR–0006 (NRC, 2003ad) and in Nuclear Assurance Corporation International (2001aa). The NRC staff finds that each of the applicant's described attributes of the nuclear material transfer reports within its Material Control and Accounting Program is the same as the subcriteria presented in YMRP Section 1.4.3, Acceptance Criterion 4, and thus the applicant's license application acceptably addresses Acceptance Criterion 4. Therefore, the NRC staff finds that, in support of a construction authorization, the applicant's described procedures for preparation of nuclear material transfer reports are adequate.

1.4.3.2.5 Findings on Material Control and Accounting

On the basis of the information the applicant provided and the preceding review, and consistent with 10 CFR 63.21(b)(4), the NRC staff finds the following.

The NRC staff finds that, in support of a construction authorization, the applicant's description of the Material Control and Accounting Program at the Yucca Mountain repository is complete and acceptably addresses each of the acceptance criteria in YMRP Section 1.4.3, and thus acceptably addresses 10 CFR 63.78. Therefore, the NRC staff finds the applicant's description of its Material Control and Accounting Program acceptable.

The NRC staff finds the applicant's commitment to submit to NRC a Material Control and Accounting Program, compliant with applicable portions of 10 CFR Part 74, no later than 180 days after issuance of a construction authorization is acceptable because the applicant submitted a description of its material control and accounting program, which acceptably addressed each of the attributes of the material control and accounting acceptance criteria, and which included a commitment to submit the Material Control and Accounting Program, compliant with applicable portions of 10 CFR Part 74, at a later date. 10 CFR 63.21(b)(4) requires DOE to only submit a description of the material control and accounting program. DOE's submission is consistent with 10 CFR Part 63, because it includes a description of the detailed material control and accounting program to meet the requirements of 10 CFR 63.78. When the detailed program DOE has committed to provide is submitted, it will describe procedures for material balance, inventory, record-keeping, and reporting (including reports of accidental criticality or loss of special nuclear material) used to demonstrate compliance with such requirements. The applicant's commitment is listed in SER Volume 1, Appendix.

1.4.4 Evaluation Findings

The NRC staff has reviewed GI and other information submitted in support of the license application and finds, with reasonable assurance, that the requirements of 10 CFR 63.21(b)(4) are satisfied with respect to a construction authorization. DOE has described a material control and accounting program that meets the requirements of 10 CFR 72.72, 72.74, 72.76, and 72.78.

1.4.5 References

DOE. 2008ab. DOE/RW–0573, "Yucca Mountain Repository License Application." Rev. 0. ML081560400. Las Vegas, Nevada: DOE, Office of Civilian Radioactive Waste Management.

NRC. 2007ae. "Geologic Repository Operations Area Security and Material Control and Accounting Requirements: Proposed Rule." *Federal Register*. Vol. 72, No. 244. pp. 72522–72562. Washington, DC: NRC.

NRC. 2003aa. NUREG–1804, "Yucca Mountain Review Plan—Final Report." Rev. 2. Washington, DC: NRC.

NRC. 2003ac. NUREG/BR–0007, "Instructions for the Preparation and Distribution of Material Status Reports (DOE/NRC Forms 742 and 742C)." Rev. 5. Washington, DC: NRC.

NRC. 2003ad. NUREG/BR–0006, "Instructions for Completing Nuclear Material Transaction Reports (DOE/NRC Forms 741 and 740M)." Rev. 5. Washington, DC: NRC.

Nuclear Assurance Corporation International. 2001aa. NMMSS Report D–24, "Personal Computer Data Input for NRC Licensees." TIC: 253781. Norcross, Georgia: Nuclear Assurance Corporation International.

USA/International Atomic Energy Agency (IAEA). 1998aa. "Protocol Additional to the Agreement Between the United States of America and the International Atomic Energy Agency for the Application of Safeguards in the United States of America."

USA/International Atomic Energy Agency (IAEA). 1977aa. "Agreement Between the United States of America and the International Atomic Energy Agency for the Application of Safeguards in the United States of America."

CHAPTER 5

1.5 Description Of Site Characterization Work

1.5.1 Introduction

This chapter of the Safety Evaluation Report (SER) evaluates the U.S. Department of Energy's (DOE or applicant) information on site characterization. The U.S. Nuclear Regulatory Commission (NRC) staff's evaluation is based on information provided in General Information (GI) Section 5 (DOE, 2009av). GI Section 5 contains the applicant's description of its site characterization activities for the Yucca Mountain site and a summary of its site characterization results. This information allows the applicant to demonstrate its understanding of what aspects of the Yucca Mountain site and its environs influence repository design and performance.

Understanding the performance of the design, in the context of the Yucca Mountain site and its environs, allows the applicant to make risk-informed, performance-based judgments regarding compliance with regulations in its Safety Analysis Report (SAR). The NRC staff evaluates the SAR in SER Volumes 2 to 5. Accordingly, the applicant's information in GI Section 5 that the NRC staff reviews in this chapter is generally informational in nature, with the more detailed technical descriptions and discussions found elsewhere in the SAR.

1.5.2 Regulatory Requirements

The requirements for description of site characterization work are in 10 CFR 63.21(b)(5), which states that the general information must include a description of work conducted to characterize the Yucca Mountain site.

The NRC staff has followed the review guidance provided in the Yucca Mountain Review Plan (YMRP) (NRC, 2003aa). As described in YMRP Section 1.5.1, no detailed technical analysis of the information addressed in YMRP Section 1.5 is required. YMRP Section 1.5.3 identifies the following two acceptance criteria that the NRC staff may consider in its evaluation:

1. The "General Information" section of the license application contains an adequate description of site characterization activities.

2. The "General Information" section of the license application contains an adequate description of site characterization results.

1.5.3 Technical Review

In SER Section 1.5.3.1 the NRC staff summarizes the applicant's information on site characterization before documenting its review following the YMRP Section 1.5.3 acceptance criteria. In SER Section 1.5.3.2, the NRC staff summarizes its review methodology. The NRC staff presents its evaluation in SER Sections 1.5.3.2.1 and 1.5.3.2.2 corresponding to the individual YMRP Section 1.5.3 acceptance criteria. The NRC staff summarizes its evaluation in SER Section 1.5.3.3.

1.5.3.1　　Summary of the DOE License Application on Site Characterization

In GI Section 5, the applicant provided a description of its site characterization work of the Yucca Mountain site. The applicant described activities that were conducted as part of its formal site characterization program for the Yucca Mountain site during the years 1988 to 2001. Also, the applicant identified that supplemental information has been collected by a variety of organizations since 2002. The applicant provided a description of its site characterization of Yucca Mountain in two sections. In GI Section 5.1 the applicant described site characterization activities performed at Yucca Mountain, and in GI Section 5.2 the applicant summarized its site characterization results. The applicant described that the results of the investigations and analyses presented in GI Sections 5.1 and 5.2 included additional studies, conducted since 2002, of the geology and hydrology of Yucca Mountain, as well as updated evaluations of volcanic and seismic information.

1.5.3.2　　NRC Staff Evaluation of Site Characterization

In the following sections the NRC staff evaluates the applicant's description of work (separate descriptions of site characterization activities and site characterizations results) conducted to characterize the Yucca Mountain site. In conducting its review, the NRC staff determines whether the applicant provided an adequate description of site characterization work in six areas: (i) geology; (ii) hydrology; (iii) geochemistry; (iv) geotechnical properties and conditions of the host rock; (v) climatology, meteorology, and other environmental sciences; and (vi) reference biosphere. The NRC staff's evaluation of the applicant's description of site characterization activities and results are provided in SER Sections 1.5.3.2.1 and 1.5.3.2.2, respectively.

1.5.3.2.1　　Site Characterization Activities

In GI Section 5.1 the applicant provided an overview of the development and implementation of the site characterization program, including information on the site characterization plan. The applicant provided (i) a description of site studies prior to development of the site characterization plan (GI Section 5.1.1); (ii) a brief description of performance allocation (GI Section 5.1.2); (iii) an overview of the site characterization plan (GI Section 5.1.3); (iv) a brief description of the role of the semiannual site characterization progress reports (GI Section 5.1.4); (v) a description of pre-site characterization and site characterization activities that resulted in site recommendation (GI Section 5.1.5); and (vi) a description of testing and monitoring activities conducted after the conclusion of site characterization (GI Section 5.1.6).

In GI Sections 5.1.5.1–5.1.5.6 and 5.1.6.1–5.1.6.6 the applicant described site characterization activities for six technical areas: (i) geology (GI Sections 5.1.5.1 and 5.1.6.1); (ii) hydrology (GI Sections 5.1.5.2 and 5.1.6.2); (iii) thermal testing and near-field geochemical characteristics (GI Sections 5.1.5.3 and 5.1.6.3); (iv) geotechnical properties (GI Sections 5.1.5.4 and 5.1.6.4); (v) meteorology and climatology (GI Sections 5.1.5.5 and 5.1.6.5); and (vi) reference biosphere (e.g., airborne mass loading; GI Sections 5.1.5.6 and 5.1.6.6).

The NRC staff's evaluation of the applicant's description of site characterization activities confirms whether the applicant provided an adequate overview of site characterization activities related to each of the six technical areas.

1.5.3.2.1.1   Geology

The NRC staff confirms that the applicant provided a description of site activities it performed to characterize the geology of the Yucca Mountain site in GI Sections 5.1.5.1 and 5.1.6.1. Also, the NRC staff confirms that the applicant provided a description of its site-specific characterization activities on the stratigraphy, tectonic setting and structural geology, seismicity monitoring, volcanism, and geomorphic processes and erosion of the Yucca Mountain site. For example, the NRC staff identifies that in GI Section 5.1.5.1 the applicant described its performance of erosion and geomorphic process studies to provide representative, site-specific data to support analyses of present and past locations and rates of erosion. Also, the NRC staff identifies that in GI Section 5.1.6.1 the applicant (i) described its high resolution aeromagnetic surveys to investigate buried geologic structures and (ii) identified that detailed ground magnetic and gravity surveys over suspected buried basaltic centers were conducted to characterize anomalies in appropriate detail for borehole siting. Therefore, on the basis of the NRC staff's review of the applicant's information in GI Sections 5.1.5.1 and 5.1.6.1, the NRC staff finds that the applicant has provided an adequate overview of its site characterization activities related to geology.

1.5.3.2.1.2   Hydrology

The NRC staff confirms that the applicant provided a description of activities it conducted to characterize the hydrology of the Yucca Mountain site in GI Sections 5.1.5.2 and 5.1.6.2. Also, the NRC staff confirms that the applicant provided a description of its site-specific characterization activities on saturated zone flow, saturated zone transport, unsaturated zone flow, unsaturated zone transport, seepage testing, and infiltration of the Yucca Mountain site. For example, the NRC staff identifies that in GI Section 5.1.5.2 the applicant described its performance of seepage and transport tests in the underground region near the Enhanced Characterization of the Repository Block Cross-Drift to investigate how water migrated through the fractured rock of Yucca Mountain. Also, the NRC staff identifies that in GI Section 5.1.6.2 the applicant described its completion of long-duration elective testing activities which evaluated unsaturated zone properties. Therefore, on the basis of the NRC staff's review of the applicant's information in GI Sections 5.1.5.2 and 5.1.6.2, the NRC staff finds that the applicant has provided an adequate overview of its site characterization activities related to hydrology.

1.5.3.2.1.3   Geochemistry

The NRC staff confirms that the applicant provided a description of activities it conducted to characterize the geochemical conditions of the Yucca Mountain site in GI Sections 5.1.3, 5.1.5.3, and 5.1.6.3. Also, the NRC staff confirms that the applicant, in GI Sections 5.1.5.3 and 5.1.6.3, described its thermal testing and near-field geochemical characterization activities. For example, the NRC staff identifies that in GI Section 5.1.5.3 the applicant described its near-field studies which examined fluid-rock interactions to investigate, through thermal tests, how changes to baseline geochemical characteristics would occur during and after the thermal period. Also, the NRC staff identifies that in GI Section 5.1.6.3 the applicant described its continued data collection and analysis from testing, which continued after site characterization associated with the Site Recommendation ended. The applicant conducted activities to evaluate (i) heat and mass loss through the Drift Scale Test bulkhead, (ii) elevated concentrations of fluoride and chloride in Drift Scale Test water samples, and (iii) the discoloration of a canister that was part of the Drift Scale Test. In addition, the NRC staff identifies that the applicant generally described other near-field investigations on (i) dust, (ii) isotopic studies of secondary minerals in the unsaturated zone, (iii) isotopic and geochemical studies of zeolitized rock to evaluate radionuclide retardation capabilities, and (iv) chemical and isotopic analyses of pore waters. Therefore, on the basis of the NRC staff's review of the applicant's information in GI Sections 5.1.3, 5.1.5.3, and 5.1.6.3, the NRC staff finds that the applicant has provided an adequate overview of its site characterization activities related to geochemistry.

1.5.3.2.1.4 Geotechnical Properties and Conditions of the Host Rock

The NRC staff confirms that the applicant provided a description of activities it performed to characterize the geotechnical properties of the Yucca Mountain site in GI Sections 5.1.5.4 and 5.1.6.4. Also, the NRC staff confirms that the applicant indicated that the studies characterized geotechnical properties that (i) could impact both proposed surface and subsurface repository structures, systems, or components considered to be important to safety until permanent closure is achieved; (ii) were required for exercising the retrieval option; and (iii) addressed postclosure needs. For example, the NRC staff identifies that in GI Section 5.1.5.4 the applicant listed the types of tests and other activities it performed in underground facilities, including its (i) Single Heater Test, (ii) Drift Scale Test, and (iii) rock mechanical field tests. The applicant performed these activities to enhance its understanding of host rock performance (e.g., rock mass quality), as well as in situ performance of installed ground control hardware. The NRC staff identifies that in GI Section 5.1.5.4 the applicant also described its completion of borehole studies, which the applicant performed to characterize site-specific seismic responses.
In addition, the NRC staff identifies that the applicant generally described its performance of activities which characterized the geotechnical properties of surface soils, alluvium, and host rock. Therefore, on the basis of the NRC staff's review of the applicant's information in GI Sections 5.1.5.4 and 5.1.6.4, the NRC staff finds that the applicant has provided an adequate overview of its site characterization activities related to geotechnical properties and conditions of the host rock.

1.5.3.2.1.5 Climatology, Meteorology, and Other Environmental Sciences

The NRC staff confirms that the applicant provided a description of activities it performed to characterize the climatology, meteorology, and other environmental sciences relevant to the Yucca Mountain site in GI Sections 5.1.5.5, 5.1.5.6, and 5.1.6.5. For example, the NRC staff identifies that in GI Section 5.1.5.5 the applicant described its Yucca Mountain Project weather stations, which the applicant established in 1985, to gather weather information, including temperature, barometric pressure, humidity, wind speed and direction, cloud cover, and rainfall quantities. The NRC staff confirms that in GI Section 5.1.5.5 the applicant described the climatology studies that it used to characterize past regional climate and to provide data for future climate evaluation. The NRC staff identifies that in GI Section 5.1.6.5 the applicant described that in addition to its climatology and meteorology studies, it conducted air quality studies, which included onsite data collection. Therefore, on the basis of the NRC staff's review of the applicant's information in GI Sections 5.1.5.5, 5.1.5.6, and 5.1.6.5, the NRC staff finds that the applicant provided an adequate overview of its site characterization activities related to climatology, meteorology, and other environmental sciences.

1.5.3.2.1.6 Reference Biosphere

The NRC staff confirms that the applicant provided a description of activities it conducted to characterize the reference biosphere in GI Sections 5.1.5.6 and 5.1.6.6. For example, the NRC staff identifies that in GI Section 5.1.5.6 the applicant described how it obtained site-specific population characteristics such as local food consumption, population distribution, and crops grown. The NRC staff identifies that the applicant described in GI Section 5.1.5.6 how it supplemented its regional and local climate and meteorological data with site-specific, air-quality data using standard U.S. Environmental Protection Agency methods for inhalable and total dust loads, and described in GI Section 5.1.6.6 how it obtained airborne mass-loading measurements in Amargosa Valley. Therefore, on the basis of the NRC staff's review of the applicant's information in GI Sections 5.1.5.6 and 5.1.6.6, the NRC staff finds that the applicant provided an adequate overview of its site characterization activities related to the reference biosphere.

1.5.3.2.1.7 Summary of NRC Staff's Evaluation of Site Characterization Activities

In SER Sections 1.5.3.2.1.1–1.5.3.2.1.6, the NRC staff evaluation finds that the applicant provided an adequate overview of its site characterization activities related to geology; hydrology; geochemistry; geotechnical properties and conditions of the host rock; climatology, meteorology, and other environmental sciences; and the reference biosphere. Because the applicant's description of the site characterization activities related to geology; hydrology; geochemistry; geotechnical properties and conditions of the host rock; climatology, meteorology, and other environmental sciences; and the reference biosphere acceptably addressed YMRP Section 1.5.3, Acceptance Criterion 1, the NRC staff finds the applicant's "General Information" section of the license application contains an adequate description of site characterization activities.

1.5.3.2.2 Site Characterization Results

In GI Section 5.2 the applicant described that results of site characterization testing are applied in screening the features, events, and processes (FEPs) to be included in the postclosure total system performance assessment, as well as the justifications for FEP exclusion. The applicant provided brief descriptions on six topical areas: geology (GI Section 5.2.1); hydrology (GI Section 5.2.2); geochemistry (GI Section 5.2.3); geotechnical properties (GI Section 5.2.4); climatology, meteorology, and other environmental factors (GI Section 5.2.5); and the reference biosphere (GI Section 5.2.6). The applicant identified in GI Section 5.2 that site FEPs that could affect repository safety, and principal results from site characterization activities, are summarized for each topical area.

The NRC staff used the guidance in YMRP Section 1.5.2, Review Method 2, in reviewing the applicant's overviews of the results of the site characterization for the six topical areas: (i) geology; (ii) hydrology; (iii) geochemistry; (iv) geotechnical properties; (v) climatology, meteorology, and other environmental sciences; and (vi) the reference biosphere. The NRC staff used the guidance [e.g., (a) through (i) for YMRP Section 1.5.2, Review Method 2, Subcriterion 1] to confirm whether the applicant provided acceptable summary descriptions that included specified details for each topical area and to confirm whether the applicant's overview for a topical area was consistent with the applicant's summaries of results for other topical areas. The NRC staff determines whether the applicant's overviews of site characterization results provided (consistent with YMRP Section 1.5.3, Acceptance Criterion 2, Subcriteria 1–3) (i) a sufficient understanding of current features and processes present in the Yucca Mountain region; (ii) adequate information for evolution of future events and processes likely to be present in the Yucca Mountain region that could affect repository safety; and (iii) a description of the reference biosphere that is consistent with present knowledge of natural processes in and around the Yucca Mountain site, including the location of the reasonably maximally exposed individual (RMEI). In the following sections, the NRC staff evaluates the applicant's description of the results of site characterization activities with respect to the six topical areas.

1.5.3.2.2.1   Geology

In reviewing the applicant's information in GI Section 5.2.1 on the results of the applicant's geological characterization activities, the NRC staff confirms that the applicant included an overview of geology that is consistent with other site characterization summaries. For example, the NRC staff confirms that the applicant's overview of faults (e.g., Solitario Canyon Fault) (in GI Section 5.2.1.3.1) and fractures (in GI Section 5.2.1.3.2) and their properties was consistent with the applicant's summary of the saturated zone conceptual model in GI Section 5.2.2.3.2.1. Also, the applicant's overview of the probabilistic volcanic hazard analysis results in GI Section 5.2.1.5.3.3 was consistent with the description in GI Section 5.2.6.2 of the biosphere pathway involving volcanic release.

The NRC staff confirms that the applicant's overview of geology included information on the nine items listed in YMRP Section 1.5.2, Review Method 2, Subcriterion 1. The NRC staff's evaluation of the applicant's information for the nine items follows sequentially.

First, the NRC staff confirms that the applicant's overview of geology in GI Section 5.2.1.1 included a description of the physiographic setting and geological features of the Yucca Mountain site. For example, the applicant described that Yucca Mountain, located in southern Nevada in the north-central part of the Basin and Range Physiographic Province, is composed of north-trending, east-dipping fault blocks of volcanic rocks.

Second, the NRC staff confirms that the applicant's overview of geology in GI Section 5.2.1.2 included a description of the principal rock units at the surface and subsurface and their stratigraphic relationships, including the stratigraphy of the various volcanic rocks and alluvial deposits and the processes that formed those rocks and deposits. For instance, the applicant identified that the proposed repository would be located in the Topopah Spring Tuff, which has a maximum thickness of about 380 m [~1,247 ft] in the area of Yucca Mountain.

Third, the NRC staff confirms that the applicant's overview of geology in GI Section 5.2.1.3 included a description of, and identified the location of potentially important stratigraphic and structural features, such as faults, fractures, and fracture (joint) sets and systems. For example, in GI Section 5.2.1.3.1, the applicant described the principal block-bounding faults, including the Solitario Canyon and Bow Ridge faults, and intrablock faults, including the Ghost Dance fault, and identified their location in GI Figure 5-35. Also, the NRC staff confirms that the applicant described in GI Section 5.2.1.3.2 the fracture systems within the volcanic rocks.

Fourth, the NRC staff confirms that the applicant's overview of geology in GI Section 5.2.1.4 identified that its description of principal geotechnical properties of the soil and rock units that are important to the design of safety-related facilities is presented in GI Section 5.2.4. The NRC staff confirms that the applicant included in GI Section 5.2.4 a description of geotechnical properties of stratigraphic units involved in operation and safety of the proposed repository, including static properties (e.g., strength and deformation characteristics), dynamic properties (as functions of shear strain), and thermal properties including thermal conductivity and thermal expansion. Also, the NRC staff confirms that the applicant identified that the geotechnical information is used to determine the potential for mechanical disruption of engineered barrier system features because of drift collapse caused by seismic events or time-dependent degradation of the rock mass.

Fifth, the NRC staff confirms that the applicant's overview of geology in GI Section 5.2.1.5 included a delineation of the proposed geologic system to be used in estimating the performance of the proposed repository. For example, the applicant indicated that Yucca Mountain is within a tectonically active portion of the North American tectonic plate that is interacting with the Pacific Plate and is characterized by the following features or processes: (i) thin crust, (ii) basaltic volcanism, (iii) high topography, (iv) block and detachment faulting, and (v) west-northwest extension.

Sixth, the NRC staff confirms that the applicant's overview of geology in GI Section 5.2.1.5 included a summary of regional geomorphic, tectonic, seismic, and volcanic models with emphasis on FEPs that may have an effect on repository operations. For example, in GI Section 5.2.1.5.1 the applicant summarized the wide variety of geologic evidence that supports a geomorphic model that includes slow rates of erosion of Yucca Mountain crests and hillslopes, and slow rates of uplift. The NRC staff also confirms that in GI Section 5.2.1.5 the applicant summarized that tectonic models for Yucca Mountain and its vicinity explain current geologic structure and are consistent with data at both local and regional scales. In GI Section 5.2.1.5.2 the applicant summarized how the seismic model for Yucca Mountain indicates that the immediate area beneath Yucca Mountain has a very low rate of seismicity but is surrounded by areas with higher rates of seismicity. The NRC staff confirms that GI Sections 5.2.1.2 and 5.2.1.5 summarized a volcanic activity model for Yucca Mountain which includes the following: (i) Yucca Mountain consists of volcanic tuffs erupted from calderas approximately 14 to 11.5 million years ago and (ii) subsequent periods of basaltic eruptions tend to be aligned along major structural trends in basins surrounding Yucca Mountain, with the last eruption approximately 80,000 years ago. In GI Section 5.2.1.6, the applicant summarized, in general terms, FEPs the applicant included in its process and abstraction models developed for the total system performance assessment and, in GI Section 5.2.1.7, the potential geologic hazards during the preclosure phase. For example, the applicant considered its conceptualization of fractured rock blocks in drifts and seismic hazard curves resulting from its seismic hazard model to estimate proposed repository performance.

Seventh, the NRC staff confirms that the applicant's overview of geology in GI Section 5.2.1.7 summarized the identification of potential geologic hazards and their related engineering measures for both the postclosure and preclosure periods. For example, the applicant identified earthquake ground motion, fault displacement, ash fall from volcanism from an external source, and rockfall under seismic loading conditions as some of the potential geologic hazards for the 100-year operational period. Also, the NRC staff confirms that the applicant described its process for identifying (screening) potential hazards for which engineering measures may be required during the operational period and that the FEP screening process addresses hazards during the postclosure period and their related engineering measures.

Eighth, the NRC staff confirms that the applicant's overview of geology in GI Section 5.2.1.5.2 included a summary evaluation of seismicity. For example, the applicant summarized the inputs to seismic hazard evaluation, the method of expert elicitation used to develop the probabilistic seismic hazard analysis, and the resulting seismic hazard curves applicable to the design and performance of surface and underground facilities at the Yucca Mountain site.

Ninth, the NRC staff confirms that the applicant's overview of geology in GI Section 5.2.1.5.3 included a summary evaluation of volcanic activity. For example, the applicant summarized the inputs to volcanic hazard evaluation, the method of expert elicitation used to develop the probabilistic volcanic hazard analysis, and the resulting volcanic hazard probability distributions applicable to assessment of repository performance (i.e., a quantitative probability distribution of the annual probability of a basaltic dike intersecting the repository footprint).

As described in the preceding nine paragraphs, the NRC staff confirms that the applicant's description of site characterization results included a sufficient understanding of current features and processes related to the geology present in the Yucca Mountain region. In addition, the NRC staff confirms that the applicant's description of site characterization results included a sufficient understanding of current features and processes related to the geology present in the Yucca Mountain region because the overview provided an adequate description of the relationship among tectonic setting (GI Section 5.2.1.5); rock and sediment types (GI Section 5.2.1); geomorphic features and processes (GI Section 5.2.1.5.1); structural framework (GI Sections 5.2.1.3.1 and 5.2.1.3.2); and seismic and volcanic features, processes, and hazards (GI Sections 5.2.1.5.2 and 5.2.1.5.3).

The NRC staff confirms that the applicant's overview provided adequate information for the evolution of future geological events and processes likely to be present in the Yucca Mountain region that could affect repository safety. For example, in GI Section 5.2.1.5.2 the applicant described the record of earthquake occurrences in the vicinity of Yucca Mountain, the location of the earthquakes, and the relevancy of previous earthquakes and their locations to the assessment of future, potentially damaging earthquakes. The NRC staff also confirms that in GI Sections 5.2.1.6 and 5.2.1.7 the applicant described geological features and processes potentially adverse to postclosure performance of the repository and associated geological data and models used for estimating repository performance.

Thus, on the basis of the NRC staff's review of the applicant's information in GI Section 5.2.1, the NRC staff finds that the applicant's description of the site characterization results regarding geology provided a sufficient understanding of current features and processes present in the Yucca Mountain region, and provided adequate information for evolution of future events and processes likely to be present in the Yucca Mountain region that could affect repository safety. Because the applicant's description of the site characterization results regarding geology acceptably addressed YMRP Section 1.5.3, Acceptance Criterion 2, Subcriteria 1–2, the NRC staff finds the applicant's description of the site characterization results regarding geology adequate.

1.5.3.2.2.2   Hydrology

In reviewing the applicant's information on GI Section 5.2.2 on the results of the applicant's hydrological characterization activities, the NRC staff confirms that the applicant included an overview of hydrology that is consistent with other site characterization summaries. For example, the NRC staff confirms that in GI Section 5.2.2.3.1.1 the applicant's overview of unsaturated zone welded and nonwelded tuffs characteristics (e.g., matrix permeability, matrix porosity, and fracture density) was consistent with the applicant's summary of radionuclide transport (e.g., velocities correlating with the fracture porosity and fracture spacing of tuffs) in GI Section 5.2.3.2.

The NRC staff also confirms that the applicant's overview of hydrology included information on the eight items listed in YMRP Section 1.5.2, Review Method 2, Subcriterion 2. The NRC staff's evaluation of the applicant's information for the eight items follows.

First, the NRC staff confirms that the applicant's overview of hydrology in GI Section 5.2.2 included a description of hydrogeologic features (e.g., aquifers and confining units), including those occurring at the location of the RMEI, with emphasis on features of known or inferred hydrologic significance. For example, in GI Sections 5.2.2 and 5.2.2.1, the applicant described five major hydrogeologic units comprising the unsaturated zone and five units, including the regional carbonate aquifer and alluvium, comprising the saturated zone, and in GI Figure 5-39, schematically showed the conceptual flow path from the repository to the accessible environment. Also, the NRC staff confirms that the applicant's description included information on hydraulic conductivity, transmissivity, porosities, permeability, and other important hydrogeologic parameters of the major hydrostratigraphic units, as appropriate. For instance, in GI Section 5.2.2.1 the applicant included information on (i) porosities and permeability in the unsaturated zone and (ii) porosities and permeability, hydraulic conductivity, and transmissivity in the saturated zone. The NRC staff also confirms that the applicant described other parameters that influence flow in the unsaturated zone (e.g., the thickness of nonwelded tuff) and other important hydrogeologic parameters in the saturated zone (e.g., geologic structures and the direction of water flow).

Second, the NRC staff confirms that the applicant's overview of hydrology in GI Sections 5.2.2.2 and 5.2.2.4 included the applicant's interpretation of the regional groundwater flow system and a discussion of the major features and controls that affect local and regional groundwater supply. For instance, the applicant described that (i) the groundwater in the region is recharged mainly by percolation of snowmelt and rain in mountains, then flows in the overall direction toward Death Valley (GI Figure 5-41), and ultimately discharges as spring flow or evapotranspiration in the southern part of the Alkali Flat–Furnace Creek groundwater basin; (ii) the average flow rate in alluvium, as defined by the specific discharge (volumetric flow rate per unit cross-sectional area) distribution in the alluvium, has been estimated to be within a range of about 1.2 to 9.4 m/yr [4 to 31 ft/yr]; and (iii) local topography, the distribution of aquifers and confining units (including alluvium), and changes in atmospheric circulation patterns are major influences on local and regional water supply.

Third, the NRC staff confirms that the applicant's overview of hydrology in GI Sections 5.2.2.1 and 5.2.2.3 included a delineation of the proposed hydrogeologic system (saturated and unsaturated) to be used in estimating the performance of the proposed repository. For example, in GI Section 5.2.2.3.1 the applicant described the unsaturated zone hydrostratigraphy and hydrology; for instance the applicant identified that the Paintbrush nonwelded unit consists of layers of predominantly nonwelded and bedded tuffs (particularly in the repository area) with high matrix porosity and low fracture frequency. The staff also confirms that in GI Section 5.2.2.3.2 the applicant described the saturated zone hydrostratigraphy and hydrology; for example the applicant identified that the Upper Volcanic Unit is the uppermost volcanic water-bearing unit of the saturated zone.

Fourth, the NRC staff confirms that the applicant's overview of hydrology included a description and discussion of local climate, including precipitation, temperature, and surface runoff. For example, in GI Section 5.2.2.4 the applicant described and discussed local climate and precipitation, noting that annual precipitation totals ranged between approximately 100 and 250 mm/yr [~ 3.9 to 9.8 in/yr]. The NRC staff also confirms that in GI Section 5.2.2.4 the applicant identified that the Yucca Mountain area has hot dry summers and, typically, dry cool winters, and the applicant identified that additional information climatic (e.g., local temperature ranges and extremes) was presented in GI Section 5.2.5. In GI Sections 5.2.2.7 and 5.2.2.9 the applicant discussed surface runoff (e.g., surface runoff results from intense rainfall from localized convective storms or from high-intensity precipitation cells within regional storm systems).

Fifth, the NRC staff confirms that the applicant's overview of hydrology in GI Section 5.2.2.5 included a discussion of groundwater quality. For example, the applicant described that groundwater in the region is marginally suitable for agriculture use and, in most respects, is suitable for potable drinking water, with some notable exceptions (e.g., fluoride and uranium content has exceeded state and federal limits in wells at Beatty and Amargosa Valley).

Sixth, the NRC staff confirms that the applicant's overview of hydrology in GI Sections 5.2.2.5 and 5.2.2.6 included a discussion of current water-use patterns, including groundwater withdrawals by aquifer source. For example, the applicant described that (i) the principal source of water for any use in the Death Valley regional flow system is groundwater; (ii) water use on the Nevada Test site was about 1 percent of the total water withdrawal in the system; and (iii) water levels in some parts of the Amargosa Farms area have been declining since the 1960s, when large-scale pumping began in the area.

Seventh, the NRC staff confirms that the applicant's overview of hydrology in GI Section 5.2.2.6 included an estimated water budget for the respective aquifer systems. For example, the applicant described water withdrawals in the entire domain of the Death Valley regional flow system, and identified total water use in Nye County use in 2000 was about 124.6 million m^3 [101,000 acre-ft], and described water levels in the immediate vicinity of Yucca Mountain (e.g., water levels near the repository are remarkably stable).

Eighth, the NRC staff confirms that the applicant's overview of hydrology included an identification of surface hydrologic features (including impoundments and stream channels, either continuous or intermittent) or other geomorphic features that could potentially affect the geologic repository operations area or safety. For example, in GI Section 5.2.2.7 the applicant identified that the only permanent surface water bodies (impoundments) in the vicinity of Yucca Mountain are those that store spring discharge in Ash Meadows and identified surface water features in the Yucca Mountain region in GI Figure 5-42. The NRC staff also confirms that in GI Section 5.2.2.7 the applicant described that ephemeral stream channels on Yucca Mountain are tributary to Fortymile Wash, and an unnamed ephemeral stream channel in Crater Flat drains the western slope of Yucca Mountain via Solitario Canyon. The NRC staff confirms that in GI Sections 5.2.2.7 and 5.2.2.9, while describing flooding and potential hydrologic hazards during the preclosure phase, the applicant identified evidence of prehistoric flooding in Coyote Wash (a tributary to Drill Hole Wash) on Yucca Mountain (i.e., sediments indicative of multiple flood events, including debris-flow deposits).

As described in the preceding eight paragraphs, the NRC staff confirms that the applicant's description of site characterization results included a sufficient understanding of current features and processes related to the hydrology present in the Yucca Mountain region. In addition, the NRC staff confirms that the applicant's description of site characterization results included a sufficient understanding of current features and processes related to the hydrology present in the Yucca Mountain region because the overview provided an adequate description of the relationship between the stratigraphy and hydrology for the unsaturated zone, saturated zone, and alluvium in GI Sections 5.2.2.1, 5.2.2.3.1, and 5.2.2.3.2. Also, the applicant provided an adequate description of surface water hydrology in GI Section 5.2.2.7 and its effect on contemporary quantity, quality, and uses of groundwater in GI Sections 5.2.2.5 and 5.2.2.6.

The NRC staff confirms that the applicant's overview provided adequate information for the evolution of future hydrological events and processes likely to be present in the Yucca Mountain region that could affect repository safety. For instance, the staff confirms that in GI Sections 5.2.2.8 and 5.2.2.9 the applicant described hydrologic features and processes used to estimate postclosure performance and potential hydrologic hazards during the preclosure period. Also, the applicant described (i) seepage into drifts (in GI Sections 5.2.2.3.1.1 and 5.2.2.8); (ii) climate change (in GI Sections 5.2.2.4 and 5.2.2.8); (iii) perched water characteristics (in GI Sections

5.2.2.1 and 5.2.2.8); (iv) water table change (in GI Sections 5.2.2.3.2.1 and 5.2.2.8); and (v) flooding (in GI Sections 5.2.2.7 and 5.2.29).

Thus, on the basis of NRC staff's review of the applicant's information in GI Section 5.2.2, the NRC staff finds that the applicant's description of the site characterization results regarding hydrology provided a sufficient understanding of current features and processes present in the Yucca Mountain region, and provided adequate information for evolution of future events and processes likely to be present in the Yucca Mountain region that could affect repository safety. Because the applicant's description of the site characterization results regarding hydrology acceptably addressed YMRP Section 1.5.3, Acceptance Criterion 2, Subcriteria 1–2, the NRC staff finds the applicant's description of the site characterization results regarding hydrology adequate.

1.5.3.2.2.3   Geochemistry

In reviewing the applicant's information in GI Section 5.2.3 on the results of the applicant's geochemical characterization activities, the NRC staff confirms that the applicant included an overview of geochemistry that is consistent with other site characterization summaries. For example, the NRC staff confirms that the applicant's overview of the chemical composition of groundwater in pores and fractures of the unsaturated zone in GI Section 5.2.3.1 is consistent with interactions between the rock units that the applicant summarized in GI Section 5.2.1.2 and the unsaturated zone groundwater flow paths summarized in GI Section 5.2.2.3.1.

The NRC staff confirms that the applicant's overview of geochemistry included information on the three factors listed in YMRP Section 1.5.2, Review Method 2, Subcriterion 3. The NRC staff's evaluation of the applicant's information for the three items follows.

First, the NRC staff confirms that the applicant's overview of geochemistry in GI Section 5.2.3 included a delineation of the proposed geochemical environment to be used in estimating repository performance. For instance, the applicant described key geochemical attributes (e.g., sorption capacities of the various geologic materials along potential flow paths) affecting the ability of the natural barrier below the repository to prevent or substantially reduce the rate of movement of radionuclides from the repository to the accessible environment.

Second, the NRC staff confirms that the applicant's overview of geochemistry in GI Section 5.2.3 included an evaluation of groundwater to determine characteristics such as water chemistry, radionuclide solubility, and radionuclide sorption capability. For example, in GI Section 5.2.3.1 the applicant described the range of water compositions in the unsaturated zone (i.e., a wide range of water compositions, ranging from calcium-chloride-type or calcium-sulfate-type waters to sodium-bicarbonate-type waters) and described how flow paths in the saturated zone derived from hydraulic analyses were compared to flow paths deduced from the hydrochemical data. The NRC staff also confirms that in GI Sections 5.2.3.1 and 5.2.3.2 the applicant described how it used the results of groundwater chemistry studies to determine solubilities of key radionuclides and sorption distribution coefficients for transport calculations (e.g., the J-13 and UE-25 p#1 waters were used in sorption experiments as end-member compositions).

Third, the NRC staff confirms that the applicant's overview of geochemistry in GI Section 5.2.3.4 included a description of the anticipated geochemical environment in the vicinity of emplaced waste packages. For example, in GI Sections 5.2.3.4.1 through 5.2.3.4.4, respectively, the applicant described (i) coupled processes in the near-field environment,

(ii) thermal-hydrologic-chemical processes, (iii) thermal hydrology in repository host rock, and (iv) in-drift chemistry.

As described in the preceding three paragraphs, the NRC staff confirms that the applicant's description of site characterization results included a sufficient understanding of current features and processes related to geochemistry present in the Yucca Mountain region. In addition, the NRC staff confirms that the applicant's description of site characterization results included a sufficient understanding of current features and processes related to geochemistry present in the Yucca Mountain region because the overview provided, in GI Sections 5.2.3.1 through 5.2.3.3 and in GI Sections 5.2.3.4.1 through 5.2.3.4.4, an adequate description of the relationship among site geochemistry, site geology, and site hydrology. Also, in GI Section 5.2.3.5 the applicant described the geochemical processes and features used to estimate postclosure performance.

The NRC staff confirms that the applicant's overview provided adequate information for evolution of future geochemical events and processes likely to be present in the Yucca Mountain region that could affect repository safety. For example, in GI Section 5.2.3.4 the applicant summarized the evolution of coupled thermal, hydrologic, and chemical processes in the near-field environment of waste packages, and in the drifts and in rocks of the repository host horizon. The NRC staff also confirms that in GI Sections 5.2.3.5 and 5.2.3.6 the applicant described geochemical features and processes relevant to the postclosure performance of the repository and the associated geochemical data and models used for estimating repository performance.

Thus, on the basis of the NRC staff's review of the applicant's information in GI Section 5.2.3, the NRC staff finds that the applicant's description of the site characterization results regarding geochemistry provided a sufficient understanding of current features and processes present in the Yucca Mountain region, and provided adequate information for evolution of future events and processes likely to be present in the Yucca Mountain region that could affect repository safety. Because the applicant's description of the site characterization results regarding geochemistry acceptably addressed YMRP Section 1.5.3, Acceptance Criterion 2, Subcriteria 1–2, the NRC staff finds the applicant's description of the site characterization results regarding geochemistry adequate.

1.5.3.2.2.4   Geotechnical Properties

In reviewing the applicant's information in GI 5.2.4 on the results of the applicant's geotechnical characterization activities, the NRC staff confirms that the applicant included an overview of geotechnical properties and conditions that was consistent with other site characterization summaries. For example, the NRC staff confirms that the applicant's overview of the types of rocks and their layering and fracturing (in GI Section 5.2.4.1) was consistent with the applicant's summary of the characteristics of the rocks in GI Section 5.2.1.2. For instance, in GI Table 5-3 the applicant presented a generalized stratigraphic column of volcanic rock in the vicinity of Yucca Mountain, which included the correlative thermal-mechanical units. Also, the applicant's overview of the soil material properties in GI Section 5.2.4.2.1 (e.g., coarse and granular alluvial deposits) was consistent with the applicant's summary of the Quaternary deposition of alluvial deposits in GI Section 5.2.1.2.

The staff confirms that the applicant's overview of geotechnical properties included information on the three items listed in YMRP Section 1.5.2, Review Method 2, Subcriterion 4. The staff's evaluation of the applicant's information for the three items follows.

First, the NRC staff confirms that the applicant's overview of geotechnical properties and conditions in GI Sections 5.2.4.1, 5.2.4.2 and 5.2.4.2.1 included a discussion of the results of site investigations necessary to characterize the engineering properties of the soils present at the site. For example, in GI Section 5.2.4.1 the applicant identified how it investigated surface and near-surface geotechnical properties and conditions at or near the location of the repository surface facilities (e.g., using boreholes); in GI Section 5.2.4.2 the applicant summarized that it used density and porosity to estimate other properties; and in GI Section 5.2.4.2.1 the applicant described that it determined total density of subsurface materials for surface facility design using gamma-gamma logging in selected boreholes.

Second, the NRC staff confirms that the applicant's overview of geotechnical properties and conditions in GI Sections 5.2.4.1, 5.2.4.2, and 5.2.4.2.2 included a discussion of the results of site investigations necessary to characterize the engineering properties of the rock types present at the site, with particular emphasis on the host rock and its immediate environs necessary for the underground excavation of the geologic repository. For example, in GI Section 5.2.4.1 the applicant identified how it examined the rock units that constitute the host rock at the repository horizon (e.g., using geophysical surveys); in GI Section 5.2.4.2 the applicant summarized that shear-wave velocity profile of the site is also a relevant input to ground motion analysis; and in GI Section 5.2.4.2.2 the applicant described that shear-wave seismic velocity profiles were sampled by spectral analysis of surface waves, downhole seismic, and vertical seismic profiling surveys.

Third, the NRC staff confirms that the applicant's overview of geotechnical properties and conditions in GI Sections 5.2.4.2.1 and 5.2.4.4 included a discussion and description of other site characterization work conducted to define the relevant geotechnical properties and anticipated response/performance of both surface and subsurface facilities. For example, the applicant described in GI Section 5.2.4.2.1 that it also applied a surface-based method of characterizing velocity (i.e., spectral analysis of surface waves) to develop seismic velocity profiles for the site. Also, the NRC staff confirms that in GI Section 5.2.4.4 the applicant discussed types of tests and other field activities performed to enhance an understanding of rock mass behavior and *in-situ* performance of ground control hardware installed to maintain the safety and stability of underground excavations (e.g., rock mass mechanical field tests).

As described in the preceding three paragraphs, the NRC staff confirms that the applicant's description of site characterization results included a sufficient understanding of current features and processes related to geotechnical properties and conditions present in the Yucca Mountain region. In addition, the NRC staff confirms that the applicant's description of site characterization results included a sufficient understanding of current features and processes related to geotechnical properties and conditions present in the Yucca Mountain region because the overview in GI Sections 5.2.4.1, 5.2.4.2.1, and 5.2.4.2.2 provided an adequate description of rock and soil properties and conditions that related them to the geology, geomorphology, and geologic hazards. Also, in GI Section 5.2.4.3 the applicant described the geotechnical features and processes used to estimate postclosure performance.

The NRC staff confirms that the applicant's overview provided adequate information for evolution of future geotechnical events and processes likely to be present in the Yucca Mountain region that could affect repository safety. For example, in GI Section 5.2.4.3 the applicant identified that mechanical properties of intact rock matrix and fractures, thermal properties, and large-scale properties of the lithophysal rock mass are important to the analysis of rockfall impacts on engineered barrier systems features, and in GI Section 5.2.4.4 the applicant described its geologic mapping from exploratory excavations (e.g., fracture geometry and characterization of the amount of lithophysal porosity, which are the primary geologic structural features affecting rock mass behavior).

Thus, on the basis of the NRC staff's review of the applicant's information in GI Section 5.2.4, the NRC staff finds that the applicant's description of the site characterization results regarding geotechnical properties and conditions provided a sufficient understanding of current features and processes present in the Yucca Mountain region, and provided adequate information for evolution of future events and processes likely to be present in the Yucca Mountain region that could affect repository safety. Because the applicant's description of the site characterization results regarding geotechnical properties and conditions acceptably addressed YMRP Section 1.5.3, Acceptance Criterion 2, Subcriteria 1–2, the NRC staff finds the applicant's description of the site characterization results regarding geotechnical properties and conditions adequate.

1.5.3.2.2.5   Climatology, Meteorology, and Other Environmental Information

In reviewing the applicant's information on the results of the applicant's site characterization, the NRC staff confirms that the applicant included in GI Sections 5.2.5 and 5.2.6.1 an overview of climatological, meteorological, and other environmental information for the site. For example, in GI Sections 5.2.5.1 through 5.2.5.6 the applicant described site characterization results on local and regional precipitation; local temperature ranges and extremes; humidity and evaporation; lightning characteristics and frequency; local wind characteristics, including tornadoes; and atmospheric stability. The NRC staff identifies that in GI Section 5.2.6.1 the applicant described, in addition to its climatologic and meteorologic results, the results of its air quality monitoring (e.g., air quality monitoring at Yucca Mountain between October 1991 and September 1995 showed levels below the applicable National Ambient Air Quality Standards for gaseous criteria pollutants and particulate matter). The NRC staff also confirms that the applicant's overview (in GI Sections 5.2.5.3 through 5.2.5.4) included a description of paleoclimate FEPs that it used as a baseline for making projections of future climate change.

The NRC staff confirms that the applicant's description of site characterization results included a sufficient understanding of current features and processes present in the Yucca Mountain region related to climatology and meteorology. For example, the applicant's overview adequately described, in GI Sections 5.2.5.1, 5.2.5.2, and 5.2.5.6, the relationships among, and the influences on, atmospheric stability, climate cycles, seasonality, and evapotranspiration. Also, in GI Section 5.2.5.5 the applicant described climate-related processes used to estimate postclosure performance.

The NRC staff confirms that the applicant's description of site characterization results provided adequate information for evolution of future events and processes likely to be present in the Yucca Mountain region that could affect repository safety. For example, in GI Sections 5.2.5.3 through 5.2.5.5 the applicant characterized future climate FEPs based upon paleoclimatological data derived from Earth orbital cycles; geochronological data from Devils Hole, Nevada (i.e., oxygen isotope record); paleoecological data from Owens Lake, California, sediments; and other environmental information.

Thus, on the basis of the NRC staff's review of the applicant's information in GI Sections 5.2.5 and 5.2.6.1, the NRC staff finds that the applicant's description of the site characterization results regarding climatology, meteorology, and other environmental information provided a sufficient understanding of current features and processes present in the Yucca Mountain region, and provided adequate information for evolution of future events and processes likely to be present in the Yucca Mountain region that could affect repository safety. Because the applicant's description of the site characterization results regarding climatology, meteorology and other environmental information acceptably addressed YMRP Section 1.5.3, Acceptance Criterion 2, Subcriteria 1–2, the NRC staff finds the applicant's description of the site characterization results regarding climatology, meteorology and other environmental information adequate.

1.5.3.2.2.6   Reference Biosphere

In reviewing the applicant's information in GI Section 5.2.6 on the results of the applicant's reference biosphere characterization activities, the NRC staff confirms that the applicant included an overview of the reference biosphere. For example, within its reference biosphere overview the applicant described in GI Section 5.2.6.1 the location and lifestyle of the RMEI and in GI Section 5.2.6.3 a summary of biosphere dose conversion factors. The NRC staff also confirms that the applicant selected biosphere pathways for dose assessments that are consistent with arid or semi-arid conditions found in a mid-latitude desert. For example, in GI Section 5.2.6.2 the applicant's description of biosphere pathways relied on present-day characteristics of the Amargosa Valley area, and the applicant described in GI Sections 5.2.5.1 the present-day climate as relatively arid. The NRC staff also confirms that in GI Sections 5.2.6.1 DOE proposed the location of the RMEI (i.e., at the southernmost boundary of the postclosure controlled area at 36° 40′ 13.6661″ North latitude at the point where the potentially contaminated groundwater enters the accessible environment) and the representative local diet and living style of the RMEI (i.e., diet and lifestyle information for the residents of the Amargosa Valley is used to characterize the RMEI).

The NRC staff confirms that in GI Sections 5.2.6.1 and 5.2.6.4 the applicant's description of site characterization results included a sufficient understanding of the current features (e.g., topography and soils) and processes related to the reference biosphere present in the Yucca Mountain region. For example, the applicant described that chemical, physical, and biological processes that are consistent with present knowledge of the conditions in the region surrounding Yucca Mountain were the fundamental elements used to develop the conceptual model of the biosphere. The NRC staff also confirms that the applicant included in GI Section 5.2.6 adequate information for evolution of future events and processes likely to be present in the Yucca Mountain region that could affect repository safety (e.g., biosphere FEPs are included in the total system performance assessment through the biosphere dose conversion factors). As described in this and the preceding paragraph, the NRC staff confirms that the applicant's description of the reference biosphere is consistent with present knowledge of natural processes in and around the Yucca Mountain site, including the location of the RMEI. Thus, on the basis of the NRC staff's review of the applicant's information in GI Section 5.2.6, the NRC staff finds that the applicant's description of the site characterization results regarding the reference biosphere (i) provided a sufficient understanding of current features and processes present in the Yucca Mountain region; (ii) provided adequate information for evolution of future events and processes likely to be present in the Yucca Mountain region that could affect repository safety; and (iii) is consistent with present knowledge of natural processes in and around the Yucca Mountain site, including the location of the RMEI. Because the applicant's description of the site characterization results regarding the reference biosphere acceptably addressed YMRP Section 1.5.3, Acceptance Criterion 2, Subcriteria 1–3, the NRC staff finds the applicant's description of the site characterization results regarding the reference biosphere adequate.

1.5.3.2.2.7   Summary of NRC Staff's Evaluation of Site Characterization Results

In SER Sections 1.5.3.2.2.1–1.5.3.2.2.6, the staff evaluation finds that the applicant provided an adequate overview of its site characterization results related to geology; hydrology; geochemistry; geotechnical properties; climatology, meteorology, and other environmental sciences; and the reference biosphere. Because the applicant's description of the site characterization results related to geology; hydrology; geochemistry; geotechnical properties; climatology, meteorology, and other environmental sciences; and the reference biosphere acceptably addressed YMRP Section 1.5.3, Acceptance Criterion 2, Subcriteria 1–3, the NRC staff finds the applicant's description of the site characterization results regarding geology; hydrology; geochemistry; geotechnical properties; climatology, meteorology, and other environmental sciences; and the reference biosphere adequate. Therefore, the NRC staff finds that the applicant's "General Information" section of the license application adequately described site characterization results.

1.5.3.2.3   Summary of NRC Staff's Evaluation of Site Characterization

On the basis of the information the applicant provided and the preceding review, the NRC staff finds that the applicant's description of site characterization work at the Yucca Mountain site is complete and acceptably addresses each of the acceptance criteria in YMRP Section 1.5.3. Therefore, the NRC staff finds that, in support of a construction authorization, the applicant's description of site characterization acceptable.

1.5.4   Evaluation Findings

The NRC staff reviewed GI Section 5 and other information submitted in support of the license application and found, with reasonable assurance, that the requirements of 10 CFR 63.21(b)(5) are satisfied with respect to a construction authorization. There is an adequate summary description of the work done to characterize the Yucca Mountain site and an adequate summary of the results from that work.

1.5.5   References

DOE. 2009av. DOE/RW–0573, "Yucca Mountain Repository License Application." Rev. 1. ML090700817. Las Vegas, Nevada: DOE, Office of Civilian Radioactive Waste Management.

NRC. 2003aa. NUREG–1804, "Yucca Mountain Review Plan—Final Report." Rev. 2. Washington, DC: NRC.

CHAPTER 6

Conclusions

The U.S. Nuclear Regulatory Commission (NRC) staff has reviewed the general information for the Yucca Mountain repository that the applicant provided in its license application.

On the basis of the information provided in the license application and the commitments specified in the Safety Evaluation Report (SER), Volume 1, Chapters 1-5 and Appendix, the NRC staff concludes that the Yucca Mountain repository meets the following requirements of 10 CFR Part 63 with respect to a construction authorization. Pursuant to 10 CFR 63.21(b), the NRC staff has made the following findings:

- 10 CFR 63.21(b)(1)—On the basis of the evaluation in SER Volume 1, Chapter 1, the NRC staff finds that the applicant included an adequate general description of the proposed geologic repository at the Yucca Mountain site, identifying the location of the geologic repository operations area, the general character of the proposed activities, and the basis for the exercise of the Commission's licensing authority.

- 10 CFR 63.21(b)(2)—On the basis of the evaluation in SER Volume 1, Chapter 2, the NRC staff finds that the applicant included proposed schedules for construction, receipt of waste, and emplacement of wastes at the proposed geologic repository operations area that are sufficiently detailed to allow NRC staff to evaluate the overall construction program for the geologic repository operations and its infrastructure.

- 10 CFR 63.21(b)(3)—On the basis of the evaluation in SER Volume 1, Chapter 3, the NRC staff finds that the applicant included an acceptable description of the detailed security measures for physical protection of high-level radioactive waste in accordance with 10 CFR 73.51 and generally described the design for physical protection, the safeguards contingency plan, the security organization personnel training and qualification plan, how the physical protection system is performance-tested to provide assurance that the system functions as intended, and how the system is tested and maintained to ensure its continued effectiveness, reliability, and availability.

- 10 CFR 63.21(b)(4)—On the basis of the evaluation in SER Volume 1, Chapter 4, the NRC staff finds that the applicant included an acceptable description of the material control and accounting program to meet the requirements of 10 CFR 63.78.

- 10 CFR 63.21(b)(5)—On the basis of the evaluation in SER Volume 1, Chapter 5, the NRC staff finds that the applicant included an adequate description of work conducted to characterize the Yucca Mountain site.

Thus, the NRC staff finds that with respect to a construction authorization DOE has adequately described the proposed geologic repository at Yucca Mountain as specified in 10 CFR 63.21(b).

CHAPTER 7

Glossary

This glossary is provided for information and is not exhaustive. The glossary provides explanations for the terms shown in italics.

absorption: The process of taking up by capillary, osmotic, solvent, or chemical action of molecules (e.g., absorption of gas by water), as distinguished from *adsorption*.

abstraction: Representation of the essential components of a *process model* into a suitable form for use in a *total system performance assessment*. Model abstraction is intended to maximize the use of limited computational resources while allowing a sufficient range of sensitivity and uncertainty analyses.

adsorption: The adhesion by chemical or physical forces of molecules or ions (as of gases or liquids) to the surface of solid bodies. For example, the transfer of solute mass, such as *radionuclides*, in *groundwater* to the solid geologic surfaces with which it comes in contact. The term *sorption* is sometimes used interchangeably with this term.

advection: The process in which solutes, particles, or molecules are transported by the motion of flowing fluid.

aging: The retention of commercial *spent nuclear fuel* on the surface in *dry storage* to reduce its thermal output as necessary to meet repository thermal management goals.

airborne mass loading: The amount of fine particulates resuspending above a surface deposit, generally expressed as mass per unit volume of air.

alluvium: Detrital (sedimentary) deposits made by flowing surface water on river beds, flood plains, and alluvial fans. It does not include subaqueous sediments of seas and lakes.

aquifer: A saturated underground geologic formation of sufficient *permeability* to transmit *groundwater* and yield water of sufficient quality and quantity to a well or spring for an intended beneficial use.

basalt: A common type of *igneous* rock that forms black, rubbly-to-smooth-surfaced lavas and black-to-red *tephra* deposits (frequently used as "lava rock" for barbecues).

borosilicate glass: A predominantly noncrystalline, relatively homogenous glass formed by melting silica and boric oxide together with other constituents such as alkali oxides. Borosilicate glass is a high-level radioactive waste material in which boron takes the place of the lime used in ordinary glass mixtures.

canister: An unshielded cylindrical metal receptacle that facilitates handling, transportation, storage, and/or disposal of high-level radioactive waste. It may serve as (i) a pour mold and container for vitrified high-level radioactive waste; (ii) a container for loose or damaged fuel rods, nonfuel components and assemblies, and other debris containing *radionuclides*; or (iii) a container that provides *radionuclide* confinement. Canisters are used in combination with specialized overpacks that provide structural support, shielding, or confinement for storage, transportation, and emplacement. Overpacks used for transportation are usually referred to as transportation *casks*; those used for emplacement in a repository are referred to as waste packages.

cask: (1) A heavily shielded container used for the dry storage or shipment (or both) of radioactive materials such as spent nuclear fuel or other high-level radioactive waste. Casks are often made from lead, concrete, or steel. Casks must meet regulatory requirements and are not intended for long-term disposal in a repository. (2) A heavily shielded container that DOE would use to transfer *canisters* between waste handling facilities at the repository.

climate: Weather conditions, including temperature, wind velocity, precipitation, and other factors, that prevail in a region.

colloid: As applied to *radionuclide* migration, colloids are large molecules or very small particles, having at least one dimension with the size range of 10^{-6} to 10^{-3} mm [10^{-8} to 10^{-5} in] that are suspended in a solvent. Colloids in *groundwater* arise from clay minerals, organic materials, or (in the context of a geologic repository) from corrosion of engineered materials.

conceptual model: A set of qualitative assumptions used to describe a system or subsystem for a given purpose. Assumptions for the *model* are compatible with one another and fit the existing data within the context of the given purpose of the *model*.

coupled processes: A representation of the interrelationships between *processes* such that the effects of variation in one process are accurately propagated among all interrelated *processes*.

criticality: The condition in which a fissile material sustains a chain reaction. It occurs when the number of neutrons present in one generation cycle equals the number generated in the previous cycle. The state is considered critical when a self-sustaining nuclear chain reaction is ongoing.

design concept: An idea of how to design and operate the aboveground and belowground portions of a repository.

diffusion: (1) The spreading or dissemination of a substance caused by concentration gradients. (2) The gradual mixing of the molecules of two or more substances because of random thermal motion.

diffusive transport: Movement of solutes because of their concentration gradient. Diffusive *transport* is the process in which substances carried in *groundwater* move through the subsurface by means of *diffusion* because of a concentration gradient.

dike: A tabular, generally vertical body of *igneous* rock that cuts across the *structure* of adjacent rocks. Dikes *transport* molten rock from depth to an erupting volcano.

direct exposure: The manner in which an individual receives dose from being in close proximity to a source of radiation. Direct exposures present an external dose *pathway*.

dispersion (hydrodynamic dispersion): (1) The tendency of a solute (substance dissolved in *groundwater*) to spread out from the path it is expected to follow if only the bulk motion of the flowing fluid were to move it. The tortuous path the solute follows through openings (pores and fractures) causes part of the dispersion effect in the rock. (2) The macroscopic outcome of the actual movement of individual solute particles through a porous medium. Dispersion dilutes solutes, including *radionuclides*, in *groundwater* and is usually an important mechanism for spreading contaminants in low flow velocities.

distribution: In a *total system performance assessment*, the overall scatter of values for a specific set of numbers (e.g., corrosion rates, values used for a particular parameter, dose results). A term used synonymously with *frequency* distribution or *probability* distribution function. Distributions have structures that are the *probability* that a given value occurs in the set.

docketing: Docketing is the acceptance of a document for placement in a docket. A docket is the information collection that constitutes the record of agency review of a license application or administrative action.

drift: From mining terminology, a horizontal underground passage. In the Yucca Mountain repository design, drifts include excavations for emplacement (*emplacement drifts*) and access (access mains).

drip shield: A metallic structure placed along the extension of the *emplacement drifts* and above the waste packages to prevent *seepage* water from directly dripping onto the waste package outer surface. The drip shield may also prevent the *drift* ceiling rocks (e.g., due to drift spallation) from falling on the waste package.

dry storage: Storage of *spent nuclear fuel* without immersion of the fuel in water for cooling or shielding; it involves encapsulating spent fuel in a steel cylinder that might be in a concrete or massive steel *cask* or structure.

dual-purpose canister: A *canister* suitable for storing (in a storage facility) and shipping (in a transportation *cask*) commercial *spent nuclear fuel* assemblies.

emplacement drift: See *drift*.

events: In a *total system performance assessment*, (i) occurrences of phenomena that have a specific starting time and, usually, a duration shorter than the time being simulated in a *model* or (ii) uncertain occurrences of phenomena that take place within a short time relative to the timeframe of the *model*.

expert elicitation: A formal, highly structured, and well-documented process whereby expert judgments, usually of multiple experts, are obtained.

fault (geologic): A planar or gently curved *fracture* across which there has been displacement parallel to the *fracture* surface.

features: Physical, chemical, thermal, or temporal characteristics of the site or potential repository system. For the purposes of screening features, *events*, and *processes* for the *total system performance assessment*, a feature is defined to be an object, structure, or condition that has a potential to affect disposal system performance.

flow: The movement of a fluid such as air, water, or magma. Flow and *transport* are *processes* that can move *radionuclides* from the proposed repository to the receptor group location.

fracture: A planar discontinuity in rock along which loss of cohesion has occurred. It is often caused by the stresses that cause folding and faulting. A fracture along which there has been displacement of the sides relative to one another is called a *fault*. A fracture along which no appreciable movement has occurred is called a joint. Fractures may act as fast paths for *groundwater* movement.

frequency: The number of occurrences of an observed or predicted event during a specific time period.

geochemical: The distribution and amounts of the chemical elements in minerals, ores, rocks, soils, water, and the atmosphere; the movement of the elements in nature on the basis of their properties.

groundwater: Water contained in pores or *fractures* in either the unsaturated or saturated zones below ground level.

high-level radioactive waste glass: A waste form produced by melting a mixture of high-level radioactive waste and components of *borosilicate glass* at a high temperature {approximately 1,100 °C [2012 °F]}.

hydrologic: Pertaining to the properties, distribution, and circulation of water on the surface of the land, in the soil and underlying rocks, and in the atmosphere.

igneous: (1) A type of rock that has formed from a molten, or partially molten, material. (2) A type of activity related to the formation and movement of molten rock either in the subsurface (intrusive) or on the surface (extrusive).

infiltration: The process of water entering the soil at the ground surface. Infiltration becomes percolation when water has moved below the depth at which evaporation or transpiration can return it to the atmosphere.

license application: An application from the U.S. Department of Energy to the U.S. Nuclear Regulatory Commission for a license to construct and operate a repository.

lithophysal: Containing lithophysae, which are holes in *tuff* and other volcanic rocks. One way lithophysae are created is by the accumulation of volcanic gases during the formation of the *tuff*.

matrix: Rock material and its pore space exclusive of *fractures*.

matrix permeability: The capability of the *matrix* to transmit fluid.

mechanical disruption: Damage to the *drip shield* or waste package because of external forces.

meteorology: The study of climatic conditions such as precipitation, wind, temperature, and relative humidity.

model: A depiction of a system, phenomenon, or process, including any hypotheses required to describe the system or explain the phenomenon or process.

near-field: The area and conditions within the potential repository including the *drifts* and waste packages and the rock immediately surrounding the *drifts*. The near-field is the region in and around the potential repository where the excavation of the repository *drifts* and the emplacement of waste have significantly impacted the natural hydrogeologic system.

parameters: Data, or values, such as those that are input to computer codes for a *total system performance assessment* calculation.

pathway: A potential route by which *radionuclides* might reach the accessible environment and pose a threat to humans. For example, *direct exposure* is a human external pathway, and inhalation and ingestion are human internal pathways.

permeability: A measure of the ease with which a fluid such as water or air moves through a rock, soil, or sediment.

porosity: The ratio of the volume occupied by openings, or voids, in a soil or rock, to the total volume of the soil or rock. Porosity is expressed as a decimal fraction or as a percentage.

probabilistic: Based on or subject to *probability*.

probability: The chance that an outcome will occur from the full set of possible outcomes. Knowledge of the exact probability of an event is usually limited by the inability to know, or compile, the complete set of possible outcomes over time or space.

probability distribution: The set of outcomes (values) and their corresponding probabilities for a random variable. See *distribution*.

processes: Phenomena and activities that have gradual, continuous interactions with the system being modeled.

process model: A depiction or representation of a process, along with any hypotheses required to describe or to explain the process.

Quaternary: The period of geologic time from about 2.6 million years ago to the present day.

Radiation Protection Program: A program for controlling and monitoring radioactive effluents and occupational radiological exposures to maintain such effluents and exposures in accordance with the requirements of 10 CFR 63.111 ("Performance objectives for the geologic repository operations area through permanent closure").

radioactivity: The property possessed by some elements (such as uranium) of spontaneously emitting energy in the form of radiation as a result of the decay (or disintegration) of an unstable atom. Radioactivity is also the term used to describe the rate at which radioactive material emits radiation.

radionuclide: An unstable isotope of an element that decays or disintegrates spontaneously, thereby emitting radiation. Approximately 5,000 natural and artificial radioisotopes have been identified.

reliability: The *probability* that the item will perform its intended function(s) under specified operating conditions for a specified period of time.

repository footprint: The outline of the outermost locations of where the waste is proposed to be emplaced in the Yucca Mountain repository.

retardation: Slowing or stopping *radionuclide* movement in *groundwater* by mechanisms that include *sorption* of *radionuclides*, *diffusion* into *rock matrix* pores and microfractures, and trapping of particles in small pore spaces or dead ends of microfractures.

risk: The *probability* that an undesirable event will occur, multiplied by the consequences of the undesirable event.

risk-informed, performance-based: A regulatory approach in which *risk* insights, engineering analysis and judgments, and performance history are used to (i) focus attention on the most important activities; (ii) establish objective criteria based on *risk* insights for evaluating performance; (iii) develop measurable or calculable *parameters* for monitoring system and licensee performance; and (iv) focus on the results as the primary basis for regulatory decisionmaking.

rockfall: The release of *fracture*-bounded blocks of rock from the *drift* wall, usually in response to an earthquake.

rock matrix: See *matrix*.

runoff: Lateral movement of water at the ground surface, such as down steep hillslopes or along channels, that is not able to infiltrate at a specified location.

seepage: The movement of *groundwater* out of *fractures* or *matrix* pores of permeable rock and into an open space in the rock. For the Yucca Mountain repository, seepage refers to water dripping into a *drift*.

seismic: Pertaining to, characteristic of, or produced by earthquakes or Earth vibrations.

seismic hazard curve: A graph showing the ground motion *parameter* of interest, such as peak ground acceleration, peak ground velocity, or spectral acceleration at a given *frequency*, plotted as a function of its annual *probability* of exceedance.

sorption: The binding, on a microscopic scale, of one substance to another. Sorption is a term that includes both *adsorption* and *absorption* and refers to the binding of dissolved *radionuclide*s onto geologic solids or waste package materials by means of close-range chemical or physical forces. Sorption is a function of the chemistry of the radionuclides, the fluid in which they are carried, and the material they encounter along the *flow* path.

spent nuclear fuel: Nuclear reactor fuel that has been used to the extent that it can no longer effectively sustain a chain reaction and that has been withdrawn from a nuclear reactor following irradiation, the constituent elements of which have not been separated by reprocessing. This fuel is more radioactive than it was before irradiation and releases significant amounts of heat from the decay of its fission product *radionuclides*.

stratigraphy: The branch of geology that deals with the definition and interpretation of rock strata; the conditions of their formation, character, arrangement, sequence, age, and distribution; and especially their correlation by the use of fossils and other means of identification. See *stratum*.

stratum: A layer of rock or soil with geologic characteristics that differ from the layers above or below it.

structure: In geology, the arrangement of the parts of geologic *features* or areas of interest such as folds or *faults*. This includes *features* such as *fractures* created by faulting and joints caused by the heating of rock. For engineering usage, see *structures, systems, and components*.

structures, systems, and components: A *structure* is an element, or a collection of elements, that provides support or enclosure, such as a building, *aging* pad, or *drip shield*. A *system* is a collection of components, such as piping; cable trays; conduits; or heating, ventilation, and air-conditioning equipment, that are assembled to perform a function. A *component* is an item of mechanical or electrical equipment, such as a *canister* transfer machine, transport and emplacement vehicle, pump, valve, or relay.

tectonic: Pertaining to geologic *features* or *events* created by deformation of the Earth's crust.

tephra: A collective term for all clastic (fragmental) materials ejected from a volcano during an eruption and transported through the air.

total system performance assessment: A *risk* assessment that quantitatively estimates how the potential Yucca Mountain repository system will perform in the future under the influence of specific *features*, *events*, and *processes*, incorporating uncertainty in the *models* and uncertainty and *variability* of the data.

transport: A process that allows substances such as contaminants, *radionuclides*, or *colloids*, to be carried in a fluid from one location to another. Transport processes include the physical mechanisms of *advection*, convection, *diffusion*, and *dispersion* and are influenced by the chemical mechanisms of *sorption*, leaching, precipitation, dissolution, and complexation.

transportation, aging, and disposal canister: A *canister* suitable for transportation from a remote location, *aging* at Yucca Mountain, and disposal at the repository.

tuff: A general term for volcanic rocks that formed from rock fragments and magma that erupted from a volcanic vent, flowed away from the vent as a suspension of solids and hot gases, or fell from the eruption cloud, and consolidated at the location of deposition. Tuff is the most abundant type of rock at the Yucca Mountain site.

unsaturated zone flow: The movement of water in the unsaturated zone, as driven by capillary, viscous, gravitational, inertial, and evaporative forces.

variability (statistical): A measure of how a quantity varies over time or space.

volcanism: Pertaining to extrusive *igneous* activity.

APPENDIX

COMMITMENTS

APPENDIX

Commitments

During the review of the Yucca Mountain License Application by the staff of the U.S. Nuclear Regulatory Commission (NRC), the U.S. Department of Energy made commitments related to the construction and operation of the geologic repository at Yucca Mountain. These commitments were identified in DOE's General Information (GI) and in a response to a NRC staff request for additional information (RAI). The following table lists a description of these commitments along with the referenced sources and implementation schedules for each.

U.S. Department of Energy Construction Authorization Commitments				
No.	Description of Commitment	Safety Evaluation Report Reference (Chapter/Section)	GI/RAI Response Reference	Implementation Schedule
1.	Update the license application (GI Figures 1-2 and 1-4) to reflect the private ownership and the correct acreage of Patent 27-83-0002	1/1.1.3.2.1	DOE, 2009*	In a future license application update
2.	Submission of Physical Protection Plan, compliant with applicable portions of 10 CFR Part 73	3/1.3.3.1, 1.3.3.2.1, and 1.3.3.2.12	GI Section 3†	No later than 180 days after NRC issues a construction authorization
3.	Submission of a Material Control and Accounting Program, compliant with applicable portions of 10 CFR Part 74	4/1.4.3.1, 1.4.3.2.1.1, and 1.4.3.2.5	GI Section 4†	No later than 180 days after NRC issues a construction authorization
*DOE. 2009. "Yucca Mountain—Response to Request for Additional Information Regarding License Application (Safety Analysis Report Section 5.8), Safety Evaluation Report Vol. 4, Chapter 2.5.8, Set 1." Letter (May 6) J.R. Williams to F. Jacobs (NRC). ML091330698. Washington, DC: DOE, Office of Technical Management. †DOE. 2008. DOE/RW–0573, "Yucca Mountain Repository License Application." Rev. 0. ML081560400. Las Vegas, Nevada: DOE, Office of Civilian Radioactive Waste Management.				

www.ingramcontent.com/pod-product-compliance
Lightning Source LLC
Chambersburg PA
CBHW081144170526
45165CB00008B/2785